High Arctic Extreme Science

Environmental Research From the Trans-Ellesmere Island Ski Expedition

Robert L. France

ISBN 978-0-9812434-0-5

First Edition

Printed in Canada
Scriptorium/Palimpsest Press is an imprint of Green Frigate Books

www.greenfrigatebooks.com
Winnipeg, Manitoba and Sheffield, Vermont

France, Robert L.
High Arctic Extreme Science: Environmental Research from the Trans-Ellesmere Island Ski Expedition.
978-0-9812434-0-5
1.Environmental Science. 2. Arctic Expedition. 3. Pollution. 4. Wildlife. 5. Garbage.

Printing this book on recycled stock has saved: 132 trees, 56,632 gallons of water, 22,785 kwh electricity, 6,231 ilbs solid waste and 12,145 lbs of greenhouse gasses.

High Arctic Extreme Science

Books written or edited by Robert L. France

Manganese in the Canadian Environment

Reflecting Heaven: Thoreau on Water

Deep Immersion: The Experience of Water

Wetland Design: Principles and Practices for Landscape Architects and Land-Use Planners

Profitably Soaked: Thoreau's Engagement with Water

Handbook of Water Sensitive Planning and Design

Road Ecology: Science and Solutions

Facilitating Watershed Management: Fostering Awareness and Stewardship

Introduction to Watershed Development: Understanding and Managing the Impacts of Sprawl

Healing Natures, Repairing Relationships: New Perspectives on Restoring Ecological Spaces and Consciousness

Wetlands of Mass Destruction: Ancient Presage for Contemporary Ecocide in Southern Iraq

Handbook of Regenerative Landscape Design

Ultreia! Onward! Progress of the Pilgrim

Restorative Redevelopment of Devastated Ecocultural Landscapes

"Dedicated with respectful memory to the diligent recordings of natural history on northern Ellesmere Island by A. W. Greely and H. W. Fielden accomplished under circumstances of such extreme hardship during the early expeditions that they can nowadays be scarcely imagined."

—from the acknowledgements in France, R.L. 1993. The Lake Hazen Trough: A late winder oasis in a polar desert. *Biol. Conserv.* 63:149-151.

"Robert L. France, Ph. D., spent a quarter of 1990 pulling a 120 kg sledge 1300 km across Ellesmere Island, the longest man-hauling expedition in the Canadian Arctic since the days of the Franklin search. He presently undertakes environmental research that does not involve sledging."

—from the author description in France, R.L. and J. M. Blais. 1998. Lead concentrations and stable isotope evidence for transpolar contamination of plants in the Canadian High Arctic. *Ambio* 27:506-508.

Table of Contents

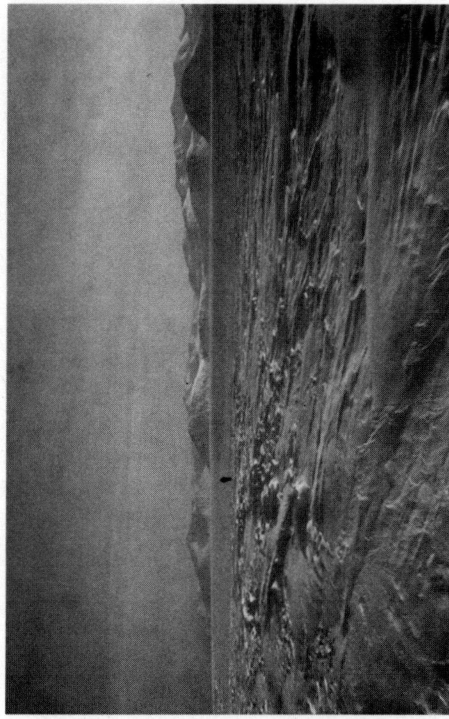

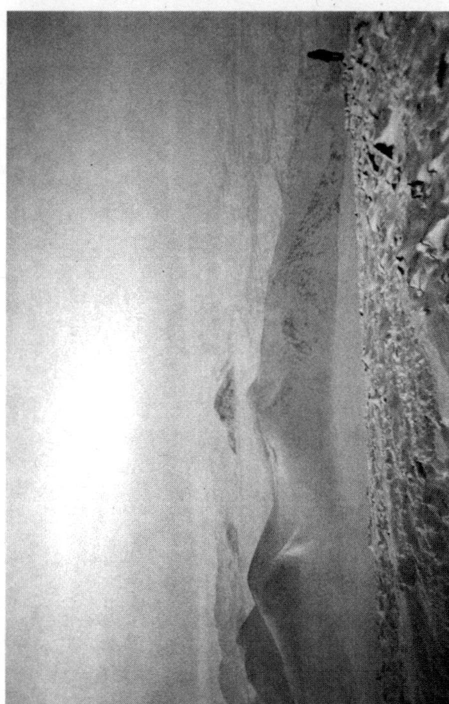

PREFACE

Return to Ellesmere Island

The production of this book complied from my research papers originating from the Trans-Ellesmere Island Ski Expedition was a welcome opportunity to relive what is likely to amount to being the grand adventure of my life. The original manuscripts for these papers were produced in a computer program of a formatting style that no longer exists commercially. It might have been possible to find a translation program to the word processing format of widespread use today, or to employ any of the several audio transcription programs currently available, or to even hire someone to retype all the text. I opted instead to take the time to do the retyping myself, and in the process, therefore, to relive the events that inspired the papers from the historic expedition of 1990. Although I was able to resist the temptation to make substantial alterations to the original texts, which had been motivated by an increased understanding of issues over the succeeding years from when they were first penned, I did however clear up egregious instances of sloppy writing and ill-formed logic. As such, the chapters in the present book represent (for better or worse) about ninety-five percent fidelity to the original texts, whether or not they had been published at the time of their writing. I have also preserved the reference to the earlier moniker of N. W. T. or Northwest Territories for the political jurisdiction of Ellesmere Island in order to remind readers that this work and writing took place in the past.

After being approached by the publisher and following my agreement to compile the papers into a single volume, my next concern was where to undertake the task. The first idea was to perhaps take advantage of a Christmas return to my childhood home of Winnipeg and there, surrounded by fields of white, blowing snow and unbelievably frigid temperatures, my arctic muse would become activated so I imagined, thereby facilitating and enriching the rewriting process. Instead, however, I opted to do this work in what at first might seem to be the most unlikely of locations in which to inspire polar thoughts: Venice. So it is in this city of onetime wayfaring and mercantile explorers (such as the Cabots, who would be the first named Europeans to visit Canada following initial contact by the vikings) where, in between giving lectures for a course at the Harvard-Ca'Foscari University Summer Program, I have sat at the computer and conceptually relived these arctic adventures in what we have called "extreme science". Often my thoughts would wander to Fra Mauro sitting in his cell on the nearby island of San Michele during the fifteenth century, putting stylus to canvas as he tried to capture the impressions of returning travelers and explorers from distant corners of the world onto his evolving map. And more than once when away from this writing table, I would experience an eerie flashback as I pulled my grocery carryall behind me, struggling up over the numerous *pontes* along the Zattere *fondamenta* with the view of the island of Giudecca shimmering in the distance across the expanse of water. At such moments, with white Palladian church facades doubling in my mind as tidewater glaciers, I would find myself wrenched back in time to manhauling my sledge over the sometimes difficult terrain of the icefoot alongside the open North Water that separated us from the Greenland shore visible in the distance.

In a way, Venice did seem appropriate for this task after all despite the occasionally nearly unbearable heat and crowds of gawking tourists. For it is in Venice more than any other major city where one can experience a deep quiet free from that mad din of automobiles that plague the rest of our metropoli, and it is Venice, arguably in the view of many, the most beautiful of all our cities, where one is repeatedly startled to attention by the luminous Sublime. And it is exactly these selfsame feelings of haunting silence and mesmerizing beauty that resonate most clearly across the years from my time spent on Ellesmere Island. Finally, it is Venice, that precarious place of alarming fragility and heightened susceptibility, that seems in some respect closest to the precious landscape and environmental concerns of the High Arctic as addressed within these pages.

And so it has been in Venice where this book was prepared, the writing broken by trips on a *vaporetto* to visit the Giardini and there to gaze up at the statue of the Italian explorer Francesco Querini, surrounded by his sled dogs, all immortalized in cold, white stone, and all staring fixedly into space toward the Lido but in fact representing their heroic last moments spent huddled together on the unforgiving polar ice.

With the collection of these papers gathered under a single cover I finally satisfy a long-held dream of producing a volume of expeditionary science in the grand tradition of those nineteenth century science-adventurers whose exploits so filled my childhood imagination. In this respect I have always been inspired by Cherry-Gerard's interlinking of the physical and scholarly pursuits of exploration. Still, as I reread through these papers, many for the first time in more than a decade, it was that which is not recorded here—in other words the story behind the research—that I have remembered most vividly about the "extreme science" of the expedition: the frostbitten fingers that struggled to work the forceps when sampling the plants, the crinkling of the plastic sample storage bags in the bitterly cold and "searching" wind, the inescapable and somewhat guilt-ridden thoughts of how the starving Greely Expedition were forced to survive by foraging upon the "rock tripe" lichen that I so casually placed in those sample bags before heading back to the tent and a hot meal after an exhausting day of sledging, the painful reality that day after day as I gathered more and more samples my sled became heavier while those of my companions were lightened by removal of food, that magical descent into the Hazen Oasis and first view of a herd of muskoxen running across and churning up the snow as if returning toward a distant place in time from which they had temporarily escaped, the delicate beauty of those "phantoms of the polar ice", the Ivory gulls, circling over the dark *nunataks* which emerged ominously from their icecap embrace, and after so much, almost at times, overpowering, silence, the arrival at a *polynya* where we would be met with a deafening cacophony of bird calls, snorting marine mammals, and wind-blown pan-ice shattering loudly as it crashed against the encircling pack-ice, and...

Well it goes on and on, and one day will hopefully be recounted. But until that time, the present compilation of the technical papers will have to suffice.

—Robert L. France
Dorsoduro, Venezia

Plate 1. Cover stories about the 1990 Trans-Ellesmere Island Ski Expedition.

Introduction

Ellesmere Island in Space and Time

Ellesmere Island lies between 76 and 83°N, is the ninth largest island in the world, contains the highest mountains in Eastern North America, is covered with expansive regions of glacial ice, contains several oases in an otherwise polar desert landscape, and is home to over two hundred species of plants, thirty species of birds, and dozen species of mammals, some of which are listed as rare or threatened (Dunn and Magor 1989). The island may have been inhabited for thousands of years by paleoeskimos and is known to have been occupied by the Thule culture people and Inuit since 1000 CE (Dick 2001). Remains of Greenland vikings have been found on the eastern coast of the island. Ellesmere Island has an extremely rich history of exploration by Europeans and southerners, developing as the gateway to the North Pole (Taylor 1955; Dick 2001). Notable nineteenth century expeditions that: (a) visited the coast of Ellesmere Island include those of Kane in 1853-55, Hayes 1860-61, Hall 1871-73, and Nares 1875-76, and (b) penetrated into the interior include those of Greely in 1881-84, Peary in 1898-1902, 1905-1906 and 1908-1909, and Sverdrup in 1898-1902. It wasn't until 1906 that the last piece of coastline was explored. Canada's northernmost permanent community, Grise Fiord, is located at the southern end of the island. Other occupation sites include the military and meteorological stations at Alert and Eureka in the northern and central parts of the island and an assortment of scientific research stations (Svoboda and Freeman 1981; Schledermann 1996; Lotz 2006). Today, the rugged and beautiful wilderness of Ellesmere Island is renowned for attracting "extreme" adventurers (e.g. Kobalenko 2002).

The Trans-Ellesmere Island Ski Expedition

Planned and organized principally by John Dunn with support from Graeme Magor and Mike Sharp, the Trans-Ellesmere Ski Island spent three months traversing roughly 1200 to 1300 km along the eastern side of Ellesmere Island during spring 1990 (Dunn 1990, 1991; Magor 1991). The route (see Appendix I) began at the northernmost tip of the island, Capes Aldrich and Columbia (the latter being the location where many expeditions for the North Pole begin), on March 30 and finished at the Inuit community Grise Fiord on July 3, with the support of two food caches established at Lake Hazen and Alexandra Fiord. Transport was by way of ski or foot and entailed each member pulling a 120 kg sled containing food and supplies. The route followed glaciated and vegetated valleys through the highest mountain ranges west of the Rockies, traveled along frozen lakes and rivers, traversed the coastal icefoot, and crossed massive inland icecaps and frozen expanses of ocean. No other people were seen from the time the pilot dropped us off at Cape Aldrich and our arrival at Grise Fiord at the end of the journey. This was likely the first north-to-south, foot-powered traverse of Ellesmere Island by people of European descent.

Review of Other Polar Expeditionary "Research"

Reading through the dozens of popular books produced by other polar expeditions, one cannot help but smile at the attempts made to create some sort of scientific credibility or environmental justification for what amounts to being in the end nothing more than a fun adventure in a cool place by a group of the "lads". It is difficult to know, of course, whether this constitutes disingenuous fraud or is an innocent manifestation of the egregious scientific illiteracy characteristic of most explorers and their sponsors. My guess is that it is the latter based on observations of one of our own expedition members who approached and conducted what he sincerely believed to be meritorious research but which was in reality of a such quality that it did not garner substantive news reportage (in contrast see Appendix II) if it was indeed published at all. Likewise, I could not find peer-reviewed articles about the stated "research" by the other adventurers despite many hours spent searching through the journals which I imagined to be likely places for publication. To be fair, it seems that the majority of these expeditions emphasized the pedagogy of their adventures, and of course it is impossible to immediately quantify the benefits of increased arctic awareness in school children that might have ensued from their following the exploits of the various expeditions.One is forced to conclude that it is a rare modern polar explorer like Fritz Koerner, arguably part of the first surface expedition to reach the North Pole in 1969, who is a real scientist of international renown and who could conduct research while engaged in the harsh rigors of polar adventure travel. This was not always the case, however. Early expeditions to Ellesmere Island before Peary, did contain extensive research programs (e.g. Fielden 1877a,b; Greely 1886).

"Extreme Science": The Expedition Research Program

My research career has been characterized by two major distinctions. First, I have done everything I could to resist the all to easy seduction of over-specialization, always keeping in mind the adage that "specialists are merely barbarians whose ignorance is not so well rounded." As a result, my research has ranged from single-celled bacteria and algae to whales, has been conducted in locations from the arctic to the tropics, and has been undertaken in a wide diversity of disciplines including statistics, chemistry, physics, meteorology, anthropology, sociology, philosophy, theology, geology, history, literature and poetry, in addition to biology and ecology (Reed 2007). I am therefore attracted by new research challenges as a way to learn about subjects about which I had hitherto little or no knowledge. And second, there has never been a sharp distinction between what I do as work and what I do as play, always keeping in mind the directive to undertake "life as a field trip". As a result, my research and travel have always been one, the advantage being that though I never stop working, this amounts to really that I never stop playing. Together, these two attributes, combined with a compelling impetus to write, has made me, with hundreds of journal articles and dozens of books, one of the most published environmental scholars of my generation (Peters 1997).

Upon first being offered a position on the Trans-Expedition Island Ski Expedition, I had to seriously consider what research I could hope to undertake before formally accepting. Never having been to the arctic before this expedition, and being at that time preoccupied with studying the effects of acid precipitation on lake ecology, I was initially at a bit of a loss about what my research program might be. Given that the lakes on Ellesmere Island would be sealed under many meters of ice at the time of our trip, and that I had no wish to carry a heavy ice-corer, I soon put aside my first wishes to undertake some form of limnological research. On the advice of my former Ph.D. supervisor Pamela Stokes at the University of Toronto, I began to investigate the possibility of collecting plants to use as biomonitors of the long-range transport, deposition, and uptake of trace contaminants. Though much recent research was emerging at the time about trans-polar air pollution in the Canadian High Arctic, most of this was based on abiotic samplings of air and snow. Little seemed to be known about how much of which contaminants were actually being incorporated into biota. Following supportive early conversations with University of Toronto's polar botanist Joseph Svoboda, and the Department of Fisheries and Oceans's Derek Muir at the Freshwater Institute where I had conducted my M.Sc. thesis research, I drafted an application and submitted it to the Canadian branch of the World Wildlife Fund, proposing to sample lichens and saxifrage in order to measure trace contaminants at the bottom of the food chain. The awarded $5,000 was enough to purchase the few sampling supplies needed, with most of the amount going toward analyses of metal, organic, and radiocesium contaminants.

Through early conversations with expedition members Graeme Magor and John Dunn, I learned that the other team member Mike Sharp was an accomplished amateur "birder", something confirmed when I finally met him when he visited Toronto before the expedition. Confident in Sharp's own ornithological capabilities and knowledge that he would also be taking an identification guide with him on the expedition, I began to read arctic wildlife papers to increase my own knowledge and identification skills.

During both these parallel avenues of researching the respective literatures, I was struck by how comparatively little was known about the environment and ecology of the High Arctic relative to temperate and tropical regions. This of course is mostly a reflection of the rarity of scientific visits to these extreme northern latitudes. Thus, I was soon satisfied that almost any serious research conducted on Ellesmere Island could become an original and worthy contribution to the literature of arctic ecology and environmental science. In the case of the present Trans-Ellesmere Island Expedition, this worthiness was all the more so given the novelty of our planned undertaking in terms of what seemed to be one of the longest latitudinal transects ever undertaken in the High Arctic. Latitude, I well knew from my previous biogeography research, was an important correlate of climate and thus expected to exert a strong influence on biology and perhaps chemical contamination as well given Ellesmere Island's nexus between North American and Eurasian sources of pollution.

In the end, my early predictions about the originality and worthiness of this research was also believed by the many scientists who reviewed the manuscripts to emerge from the expedition. Indeed, in one case a wildlife paper was republished in a collection representing the best conservation biology work of the 1990s, and in another case a contamination paper was selected as the featured cover story for that issue of the prominent journal in which it occurred. Many others have also been interested in these findings. In the years since publication of the papers, I have received hundreds of reprint requests from environmental scientists and scholars around the world, something which remarkably continues today (hence the present justification to collect the papers and republish them together under a common cover). On the other hand, it is worth noting that while on the expedition I also conducted a good deal of other ecological, geological, and biogeochemical research which, however, has not been published due to my concerns about its quality.

Three major difficulties also affected the production and dissemination of the research from the Trans-Ellesmere Island Ski Expedition: 1) saxifrage and lichen samples collected from several locations at the start of the journey, which would have provided the northernmost biomonitoring measurements ever collected, were lost following their drop-off at the Lake Hazen food cache and their hoped-for transport and storage back at the Canadian Polar Shelf building in Resolute; 2) plant samples collected along the entire traverse of Ellesmere Island which were sent to the laboratory of P. Stokes at Trent University for what would have been the most comprehensive and northernmost mercury analyses ever undertaken, were also lost; and 3) attempts by G. Magor to prevent publication of several papers from the expedition because, in his mind, they were of poor scientific quality and thus were an embarrassment to the reputation of the Expedition, caused one editor to reject a manuscript that had originally been submitted there (but which was subsequently published without difficulty elsewhere).

Collation of Published and New Findings

The original manuscripts were written while I was a post-doctoral fellow at McGill University in the following chronological sequence, beginning first with the survey papers that did not require laboratory analysis: Chapter 12 on the Rough-legged hawk submitted to *The Canadian Field Naturalist* on December 3, 1990 (i.e. 5 months after returning from the expedition), accepted on February 25, 1992, and published as 106:511-512 in 1992; Chapter 11 on Ivory gulls submitted to *Arctic* on December 10, 1990, accepted on December 12, 1991, and published as 45:306-307 in 1992; Chapter 13 on aerial mobbing submitted to the *Journal of Raptor Research* on February 12, 1991, accepted on December 10, 1991, and published as 26:269 in 1992; Chapter 7 on the Lake Hazen oasis submitted to *Biological Conservation* on April 12, 1991, accepted February 5, 1992, and published as 63:149-151 in 1993; Chapter 8 on polynyas submitted to *Conservation Biology* on April 22, 1991, accepted on October 24, 1991, and published as 6:442-446 in 1992 and again in Ehrenfield (1995); Chapter 9 on avian species richness submitted to *Polarforschung* on May 3, 1991, and published as 61:147-152 in 1992; Chapter 14 on garbage submitted to *Nature* on October 12, 1991, and published as 355:504 in 1992; and then following with the trace contaminant papers: Chapter 1 on radiocesium submitted to the *Canadian Journal of Botany* on August 7, 1992, and published as 71:708-711 in 1993; Chapter 5 on lead in lichens submitted to *Water,*

Air and Soil Pollution on May 20, 1994, accepted on April 22, 1995, and published as 90:469-474 in 1996; Chapter 3 on organochlorine distribution submitted to *Environmental Science and Technology* on June 17, 1996, accepted February 21, 1997, and published as 31:1879-1882 in 1997; Chapter 4 on PCBs submitted to the *Bulletin of Environmental Contamination and Toxicology* on January 10, 1997, and published as 59:76-82 in 1997; and Chapter 6 on lead stable isotopes submitted to *Ambio* on January 30, 1997, accepted on August 28, 1997, and published as 27:506-508 in 1998 as the cover story. All these journals are thanked for permitting republication of these papers in the present volume.

Four other manuscripts are published in the present book for the first time: Chapter 2 on the latitudinal distribution of organochlorines was submitted to and actually accepted by *Chemosphere* but they required a reformatting of the text which would have necessitated retyping the entire paper, something that at the time (in 1998) of starting my new position at Harvard University I could not find the time for; Chapter 10 on the ice cap survey was submitted as a note to *Polar Record* in 1994 but unlike all other professional journals they never communicated back despite a subsequent inquiry; Chapter 15 consists of two letters to editors that were written in the spring of 1992 in response to several inflammatory writings that appeared in print by John England, who had been "outed" by a newspaper reporter as being a major contributor to the garbage found by our expedition (unfairly in the minds of many, these responses of mine were not published since according to the editors "I had already had my say"; why this was imagined to be so since I had gentlemanly not mentioned the offending culprit by name, puzzled many people from around the world who wrote me letters of inquiry and support at the time); and Chapter 16 on the expanded garbage survey was submitted to the *Canadian Geographic Magazine* in late 1991 but possibly because of the provocative nature of the article and the magazine's environmental timidity at that time, they declined to publish. And though this article was latter selected for publication in 1992 by *The Ecologist*, it was never actually published there to my knowledge.

Acknowledgements

Expedition members

The Trans-Ellesmere Island Ski Expedition was conceived by J. Dunn and was organized by himself and G. Magor with assistance from M. Sharp. J. Dunn is thanked for his vision, leadership, and purposeful drive that kept us moving forward day after day, sometimes under conditions so arduous that still now, two decades later, I can scarcely imagine how we all managed to do so. G. Magor is thanked for his supportive companionship during the expedition. And M. Sharp is especially thanked for his enthusiastic humour and interest in my research program, the latter to the point of his willingness to undertake keeping the important tally of all the bird sightings made along the route.

Those armchair readers of the polar and mountaineering literatures will not be surprised to learn that, like most expeditions, there was a certain degree of tension among the participants. This is very much an important part of the telling of the overall story, but one whose time and place does not fit with the present technical compilation of the expeditionary research. It is enough here to mention that for me at least the expedition was similar to the situation of no doubt many university students who have had to temporarily share accommodation; i.e. once that cohabitation is over, individuals go their separate ways often with little or perhaps no interactions with one another despite having shared mutually important experiences.

In this light, it is worth recounting that I received several letters from M. Sharp after the expedition but then lost track of his whereabouts (addresses are hard to come by for someone who was then spending more than half of his year living in snowbound tents near both the South and North Poles!). I saw J. Dunn give two public lectures about the expedition: one in Nanaimo in which he knew I was in attendance and was very cordial, and another in Ottawa in which my presence was unknown to him and in consequence my inexperienced role in the expedition was made to be one of humour for the audience. I saw G. Magor once after the expedition when we spent an evening together framing the expeditionary photos that were to be given to sponsors. My only other interaction with him turned out to be surprisingly hostile (on his part) when he attempted to prevent publication of some of the research papers by taking it upon himself to write to the journals to whom I had submitted the manuscripts (with copies kindly provided to him for his own interest), arguing that in his opinion none of the work had any professional merit. This "bizarre behavior", as one journal editor phrased it in a letter to me, luckily did not dissuade either him or, with a sole exception, any of the other editors from publishing the manuscripts. The fact that these findings were ultimately published in journals of recognized international quality following the customary peer-review process, and indeed that one of the papers was later republished in a collection representing the most important conservation biology papers of the 1990s, indicates that the inexplicable attempt to muzzle the expedition's science was unjustified. In other letters to me G. Magor continued a litany of complaints about how I had let the expedition down, how my actions were a general disservice to its memory in particular and it seems possibly to humanity in general, and so on. Sad, perhaps, but not unexpected given the sort of competitive mentality that psychologists tell us is characteristic of many ultra-marathoners and extreme outdoor adventurers, and how it affects their mindsets and interactions with the world and other people.

CO-AUTHORS

The issue of co-authorship in the publication of technical, peer-reviewed journal papers is a complicated and sometimes contentious one, with some primary researchers believing that all individuals who participated in any way in the production of the research should be given credit as co-authors whereas others (the majority) insisting that only those colleagues who helped to formulate the research program and actually contributed with "pen to paper" in the final writing stage are meritorious as being co-authors, everyone else who may have provided technical assistance being mentioned to the "acknowledgements" section. Throughout my career I have tended to be generous in the inclusion of individuals as co-authors. That said, it is important to clarify the role of others in the papers complied in this volume.

All the texts were written by myself with the few following exceptions: parts of the radiocesium counting methodology in Chapter 1 by H. Taylor and J. Svoboda, parts of the organochlorine determination methodology in Chapter 3 by D. Muir and M. Segstro, the lead determination methodology in Chapter 5 by M. Coquery, and parts of the lead isotope methodology in Chapter 6 by J. Blais. It is important to note that J. Sovboda, D. Muir and J. Blais carefully read through the manuscript drafts with their respective methodological contributions and made valuable suggestions on the overall texts. The fact that my initial drafts were very close to the final published texts is a reflection of the clarity of these scientists' previously published papers with which I became intimately familiar and used as the basis for my own writings on these subjects.

To all my coauthors, then—M. Sharp, D. Muir, J. Blais, J. Svoboda, H. Taylor, M. Coquery, M. Segstro—I offer my warmest thanks for their contributing efforts. In particular, conversations with J. Svoboda, D. Muir and J. Blais during my writing of these papers was instrumental to the ideas contained therein.

Logistic support

Shipment of plant samples from the Arctic was provided by the Atmospheric Environment Service of Canada through D. Whelpdale. The lichen samples from Sverdrup Pass analyzed for lead in Chapter 5 were collected and identified by J. Svoboda. B. Billeck collected saxifrage samples from Axel Heiberg Island that were analyzed for organochlorines in Chapter 3, which along with all the OC samples, were confirmed via high-resolution mass spectrometry by G. Stern.

Funding and sponsors

The Trans-Ellesmere Island Ski Expedition was sponsored by the Royal Canadian Geographical Society. Funding for the trace contaminants research program was supported by the World Wildlife Fund Canada and was supplemented by the operating grants of J. Svoboda, H. Taylor, D. Muir, P. Stokes, J. Kalff, and R. Peters. Support for original manuscript preparation was provided by a NSERC Strategic Grant to R. France and R. Peters for ecoforestry research. Expedition logistic support was provided by Asnes, J. Ingle Insurance, *The Calgary Herald*, Bolle, Paris Gloves, Coll-tex, Stillongs, Harvest Foodworks, MSR, Kauffman Footwear, North Face, Lowe, Leki, C.E. Jamieson, Mountain Equipment Co-op, and the Canadian Himalayan Foundation, Calgary.

Plate 2. Arctic haze observed at northern Ellesmere Island at the start of the expedition.

PART I:
TRACE CONTAMINATES

CHAPTER 1

Latitudinal Distribution of Cesium-137 Fallout in 1990 on Saxifraga oppositofolia from Ellesmere Island, Canada*

***Adapted from *Can. J. Bot.* 71:708-711; Abstract in Appendix II; Coauthors are J. Svoboda and H.W. Taylor**

INTRODUCTION

During the 1950s and 1960s, considerable amounts of fission products were injected into the atmosphere through nuclear weapons testing. Although the bulk of these radionuclides were deposited in temperate regions of the Northern Hemisphere (Hardy et al. 1968), owing to long half-lives for some of the fission products (e.g. 30.17 yr for ^{137}Cesium) and relative inactivity of the arctic environment, the slow-growing tundra plants are a major biological carrier for radioactive fallout (Taylor et al. 1985). Although ^{137}Cesium (Cs) can sometimes be incorporated into plants through uptake from the soil as a surrogate for K (Russell 1966; Mattsson and Liken 1975), direct adsorption and foliar sorption is the principal means of accumulation by arctic vegetation (Richard et al. 1965; Hanson 1968; Hutchison-Benson et al. 1985; Baskaran et al. 1991).

Despite much diminished concentrations of ^{137}Cs in precipitation over northern Canada since the period of peak fallout during the early 1960s (Bird 1968; Barrie et al. 1992), later atmospheric explosions by the Chinese and French, as well as accidents in Soviet reactors, have all been detected in the arctic vegetation of Canada (Koranda and Martin 1971; Svoboda and Taylor 1979; Taylor et al. 1979, 1988). Concern remains high about these issues because of the tenacious ability of artic plants to retain radionuclides and ^{90}Sr, ^{137}Cs is taken up into muscle tissue and therefore has a high potential for biomagnification (Mohindra 1967), increasing by a factor of two for each successive trophic level. Human concentrations in caribou consumers can be over 100 times higher than those for people residing in temperate regions (Hanson 1967, 1968; Bird 1968).

Relatively few measurements have been made of radioactive fallout in the High Arctic north of 76° latitude. Cesium-137 has been detected in precipitation (Barrie et al. 1990, ice cores (Ambach and Muller 1980), and soil samples (Hutchison-Benson et al. 1985) from Ellesmere Island, Canada. Similarly, ^{137}Cs was measured in 80 lichen samples from 27 sites surrounding Thule, Greenland (Hanson 1971), 21 samples of lichen and vascular cushion plants from 4 sites along the length of Ellesmere Island (Hutchison-Benson et al. 1985), and 7 lichen samples from one site on central Ellesmere Island (Taylor et al. 1988).

The purpose of the present study was to determine the concentration of ^{137}Cs in purple saxifrage (*Saxifraga oppositifolia*) from 10 sites spanning 6° latitude along the length of Ellesmere Island. Despite the numerous radiological studies that were conducted in artic regions, only one previously investigated the latitudinal distribution of fallout products (Hutchison-Benson et al. 1985; Taylor et al. 1985).

Methods

Saxifrage samples were collected from May 2 to June 29, 1990, as part of the Trans-Ellesmere Island Ski Expedition (Figure 1.1). Although some earlier work had indicated that lichens and mosses contain considerably more radioactivity then vascular plants (e.g. Gorham 1959; Rickard et al. 1965), other research showed that the persistent aerial parts of cushion plants may contain levels of ^{137}Cs comparable to lichens in the same arctic habitat (Koranda and Martin 1971; Svoboda and Taylor 1979). Four to 5 samples of nonflowering clumps of saxifrage were gathered from areas ranging between 10 and 50 m^2 for each site. Both live and dead material (which can contribute up to three-quarters of the biomass of saxifrage; Stewart 1984) were sampled as free of dirt as possible. Because previous research showed that the highest concentrations of ^{137}Cs occurred in vegetation growing on exposed windswept ridges that are free of snow for much of the year (Hanson 1967, 1968; Hutchison-Benson et al. 1985), such locations were selected for the present survey. In addition to latitude and longitude, the following information was documented for each sampling site: distance for the coast, elevation, substrate typology, site topography, immediate surface moisture conditions, windbreaks or obstructions within 100 m, surrounding area characteristics (1000 m), and site orientation. Samples were stored in sealed, reinforced paper bags from which they were removed for counting during 1991. Seven lichen samples were also collected: the species *Certaria nivalis* and *Alectoria ochroleura* from Sverdrup Pass (~ 79° N) and 5 replicates of *Umbilicaria muchlenbergii* from King Edward Point (~ 76° N).

Proximal samples of saxifrage were pooled to obtain two replicates averaging 39.6 ±9.5 (SD) g dry wt per field site. The samples were counted with a λ-ray spectrophotometer that used an intrinsic Ge counter (26% efficiency, 2.5 keV resolution). Each sample was counted for 24 hr. The counting rate (counts per minute per kg dry wt) was determined for each sample by integrating the 662 keV λ-ray line due to ^{137}Cs. Each λ-ray spectrum was searched for the presence of ^{134}Cs in the sample. The overall average error in the counting rates was about 12%.

Results and Discussion

A progressive decrease was observed in the ^{137}Cs activity of saxifrage with increasing latitude north of 77° 42' on Ellesmere Island in 1990 (Figure 1.2a). Cesium-137 levels in saxifrage at the two sites south of this pronounced maximum (0.07 Bq/g dry wt) were also reduced. The ^{137}Cs contents for the two lichen species from Sverdrup Pass (0.19 and 0.23 Bq/g dry wt for C. *nivalis* and *A. ochroleuca*, respectively) and for *U. muchlenbergii* from King

Edward Point (0.13 + 0.03 Bq/g dry wt) were considerably higher than those for nearby saxifrage.

The latitudinal cline in ^{137}Cs activity in Ellesmere Island saxifrage supports results found previously over a much greater latitudinal range for this species and for other vascular cushion plants and lichens (Hutchison-Benson et al. 1985). Linear decreases in plant ^{137}Cs at higher latitudes (Figure 1.2a), reflect progressively lower levels of fallout that occurred there (Hardy et al 1968; Hutchison-Benson et al. 1985). Interestingly, Hutchison-Benson et al. (1985) found that the highest levels of ^{137}Cs in vegetation occurred between 60 and 70° N. Lower ^{137}Cs values in plants at more southern regions results from higher precipitation and temperatures and therefore from correspondingly higher biological turnover rates (Taylor et al. 1985). Alternatively, the low ^{137}Cs activity for the two most southerly sites on Ellesmere Island may simply reflect different conditions of precipitation and soil drainage. Mesic habitats have longer lasting snow cover that will reduce the time period for direct deposition, as well as enabling the plants to grow faster and thereby dilute the burden of ^{137}Cs over larger biomass (Hanson and Eberhardt 1971; Hutchison-Benson et al. 1985).

Fallout particles adhering to the dead debris of vascular plants may reflect earlier deposition (Mattsson and Liden 1975; Svoboda and Taylor 1979) and therefore contribute to burdens that may exceed those of nearby lichens (Hanson 1971). In the present case, however, the activity of ^{137}Cs in slower growing lichens is not as diluted over time as in saxifrage and therefore remains higher (cf. Hutchison-Benson et al. 1985).

The collection of *S. oppositofolia* samples made it possible to obtain an estimate of the average value of the effective half-life (T_{eff}) over northern Ellesmere Island (78.8 – 82.0°N). In Figure 1.2a, along with the 1990 measurements, we include the 1979-1980 data collected for saxifrage (Hutchison-Benson et al. 1985). To obtain an estimate of the effective half-life we averaged the 1979-1980 data over the latitudinal range 78 – 82° N and the 1990 data over the range 78.8 – 81.0°N. Assuming an exponential decay over this decade, we determined the T_{eff} to be 6.2 ± 1.0 yr. To put this result in perspective, we show in Figure 1.2b the data that were used to deduce a possible empirical representation of T_{eff} as a function of latitude (after Taylor et al. 1985). The deposition data of Hardy et al. (1968) were used together with the 1979-1980 load data (Hutchison-Benson et al. 1985) to calculate T_{eff} versus latitude, assuming an empirical formula

$$T_{eff} = a + b\,\lambda(-50)$$

where a and b are empirical constants, and λ is the latitude in degrees (Taylor et al. 1985). The agreement between the 1990 datum point and the broken line from 1979-1980 (Figure 1.2b) should not be regarded as a verification of the assumed T_{eff} formula. However, the empirical work of 1979-1980 does seem to have predicted the approximate magnitude of T_{eff} for at least this one species of plant.

Although trace amounts of ^{134}Cs (half-life of 2.1 yr) have been detected in the atmosphere over northern Ellesmere Island following the July 1986 Chernobyl accident (Barrie et al. 1992), the bulk of the deposition occurred at lower latitudes in North America (Joshi 1987; Taylor et al. 1988; Baskaran et al. 1991). If it is assumed that some of the airborne activity (Barrie et al. 1992) reached the ground of Ellesmere Island, analysis of our most contaminated samples suggests that less than 30% of the minuscule load of ^{137}Cs might be ascribed to the Chernobyl accident.

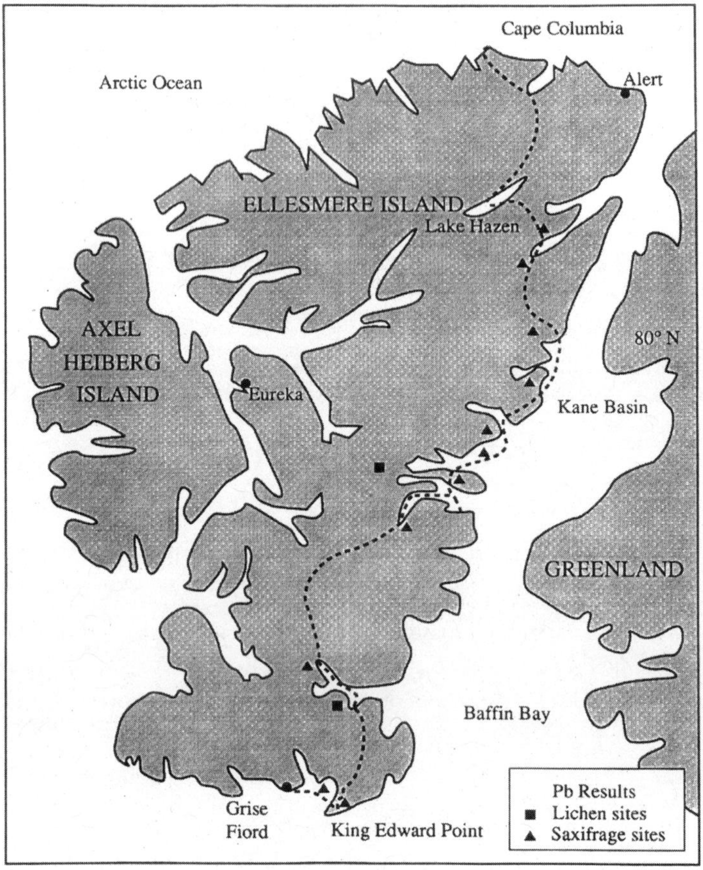

Figure 1.1. Approximate location of sites sampled for ^{137}Cs from May 2 to June 29, 1990 during the Trans-Ellesmere Island Expedition (broken line). Site identifications are from north to south (see Figure 1.2 and text for precise latitudes) are as follows: Dodge Valley, Archer Fiord, Radmore Harbour, Joiner Bay, Norman Lockyer Island, Sverdrup Pass, Flagler Bay Polynya, Jokel Fiord, Makinson Inlet (head), Jones Sound, Craig Harbour, and King Edward Point.

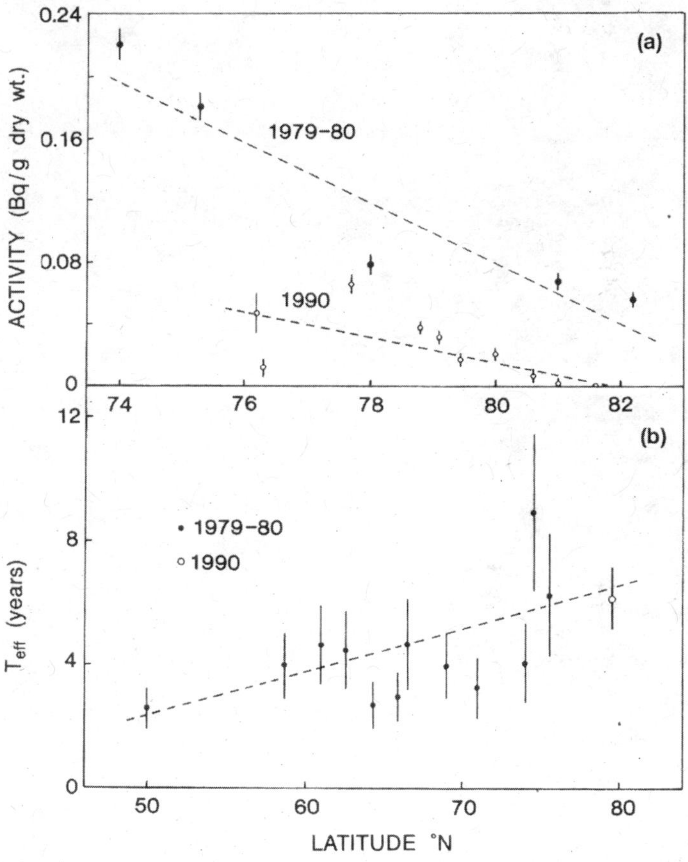

Figure 1.2. (a) Latitudinal distribution of ^{137}Cs activity in Saxifraga oppositofolia *(average + range) from Ellesmere Island in 1990 (open circles) and in 1979-1980 (solid circles) from Taylor et al. (1985). Variation of the empirically estimated effective half-life (T_{eff}) of saxifrage with latitude (average ± 90% CI) from Taylor et al. (1985) (closed circles). Open circle denotes actual measured value basded on the present 1990 data at 80°N.*

Plate 3. Dormant and flowering saxifrage from central and southern Ellesmere Island.

Chapter 2

Persistent Organochlorine and Current Use Pesticides in Lichens on Ellesmere Island in the Canadian High Arctic*

***Abstract in Appendix II**

Introduction

Atmospheric monitoring has revealed the presence of organochlorine (OC) contaminants in both the North American Arctic in general (e.g. Barrie et al. 1992; Oehme 1991), as well as the Canadian High Arctic in particular (e.g. Hoff and Chan 1986; Patton et al. 1989, 1991; Hargrave et al. 1988; Gregor and Gummer 1989; McNeely and Gummer 1984). Although OC residues have been detected in tissues of arctic animals (e.g. Norstom et al. 1988; Muir et al. 1988, 1990; Addison and Zinck 1986; Cameron and Weis 1993; Walker 1977), only Bidleman et al. (1989) have made such measurements for fauna resident in the High Arctic north of 75°N. Because of wide organismal distributions (e.g. Norstrom et al. 1988), variable metabolic processes, and complex foodweb interactions (e.g. Kidd et al. 1995), it may often be difficult to infer localized patterns in OC deposition from animal concentrations alone. Use of plants can circumvent these difficulties, and in particular, lichens, because of their absence of a cuticle, lack of internal transport mechanisms, and dependence on an atmospheric supply of nutrients, function as ideal biomonitors of site-specific cumulative OC deposition (e.g. Bacci et al. 1986; Focardi et al. 1991; Thomas 1986; Villenuve et al. 1988; Villeneuve and Holm 1984; Muir et al. 1993). The purpose of the present study was to determine, for the first time, OC concentrations in lichens from Ellesmere Island in the High Arctic, and to compare these measurements with those determined previously for lichens collected from more southerly locations.

Materials and Methods

Lichens (*Umbilicaria muchlenbergii*) were collected in June and July 1990 from two sites on southern Ellesmere Island in the Canadian High Arctic: Makinson Inlet near Bowman Island at about 77°N, and King Edward Point on the southernmost tip of Ellesmere at about 76°N (see map with sampling locations in Chapter 1). Both sites were exposed windswept cliffs from which material from horizontal ledges and vertical faces was collected. Three replicates were obtained from each study site. Samples were stored in aluminum foil-lined envelopes sealed within plastic bags from which lichens were later removed for analysis following thawing. Triplicate samples were analyzed for OCs using procedures described in Muir et al. (1988, 1990, 1993) and Kidd et al. (1995). Lichens from King Edward Point were also collected for determinations of tissue cesium (Chapter 1) and lead (Chapter 5) contents.

RESULTS

The polychloranated camphene toxaphene was by far the most abundant OC measured in these lichen samples (17-24 ng/g dry wt), representing 59% and 70% of the total accumulated OCs at the Makinson Inlet and King Edward Point sites, respectively (Table 2.1). S-PCB (sum of polychloronated biphenyls) concentrations of 3-4 ng/g dry wt were the second most abundant OC. The mono/dichlorobiphenyls and congener-1 were the most common forms of PCBs detected in the lichens, followed by the congeners-118, 101 and 153 (hexachlorobiphenyls). S-HCH (sum of hexachlorocyclohexanes) and S-CHLOR (sum of chlordanes) levels of about 2 ng/g dry wt were the next most abundant OC measured in the lichens. Cis-chlordane, trans-nonachlor, cis-nonachlor and heptachlor epoxide were the most common chlordane isomers detected. Alpha-HCH exceeded g-HCH by a ratio of 5.3:1 in lichens from both sampling sites. S-DDT (sum of DDT isomers) and S-CBz (sum of chlorobenzenes) displayed lichen concentrations of about 1 ng/g dry wt. The major form of DDT present in lichens was p,p'-DDE which represented about 35% of the total. Hexachlorobenzene (HCBz) was the most common chlorobenzene isomer. The other OCs: dieldrin, endrin, endosulfan, PCA, MECOL, S-VER and TRIFLU were found to occur in lichen concentrations of less than 1 ng/g dry wt (Table 2.1).

DISCUSSION

As Gregor and Gummer (1989) stated: "Despite the expanding information base that the Arctic atmosphere contained OCs, there is only limited evidence of their deposition, accumulation, and subsequent inclusion into foodwebs". The present results comprise the first record of OC contamination into terrestrial plants in the Canadian Arctic, and together with Bidleman et al.'s data on zooplankton and amphipods obtained from near Axel Heiberg Island (1989), represent conclusive evidence that the emission of OCs to the troposphere at southern latitudes is substantial enough for these contaminants to be transported to and detected in biota residing in the High Arctic (> 75° N).

The OCs a-HCH, HCB, g-HCH, DDE and DDT in lichens from Ellesmere Island are substantially lower in concentration than those for lichens from southern Europe (Villeneuve et al. 1988; Gaggi et al. 1985) and polluted northern Europe (Carlberg et al. 1983), marginally lower in concentration than those measured for lichens from southern Canada (Muir et al. 1993), and comparable in concentration to those residues in lichens collected from other remote polar sites in Spitsbergen (Calamari et al. 1991, 1994) and Antarctica (Bacci et al. 1986). Northward latitudinal declines in OC contamination have been measured in freshwater fish (Muir et al. 1990), polar bears (Norstrom et al. 1988), seals

(Luckas et al. 1990), in addition to vascular plants (Calamari et al. 1991), and reflect temperature related processes of fractionation and condensation (Wania and Mackay 1993). With the exception of the two most polluted sites, HCB concentrations show little difference between polar and temperate lichens. Calamari et al. (1991) actually found higher HCB residues in polar compared to temperate vascular plants and suggested that this was due to the further northward transport of this high volatility OC (cf. Wania and Mackay 1993).

Calamari et al. (1994) believed that the relative (i.e. mean normalized) composition of selected OCs (a-HCH, HCB, g-HCH, DDE and DDT) in plants provided a convenient "fingerprint" of regional differences in pesticide use, climate and deposition patterns. Wania and Makay (1993) summarized the extant data on animal OC residues and found that the OC spectrum often shifts from a pattern with comparable concentrations of PCBs and HCHs in temperate regions to one dominated by HCHs and HCB in the Arctic. Synthesis of global lichen data indicate that the most overt difference between southern-polluted, temperate non-polluted, and polar samples was, in support of animal data, an increasing proportional representation of HCB in the more remote locations.

Muir et al. (1990) considered that the presence of chlordane-related compounds at levels comparable to S-PCB and S-DDT is a situation unique to polar regions. The present S-CHLOR:S-PCB ratios of 0.4 and 0.5 for lichens are much higher than those measured for southern biota, suggesting, therefore, a proportionally greater input of chlordane to the Arctic.

High ratios of a-HCH to g-HCH for Ellesmere Island lichens (5.3:1) were comparable to those ratios recorded for southern Canadian lichens (ca. 6.5:1), but much larger than those ratios for lichens from European sites (ca. 1.6:1). Such high ratios indicate the absence of influxes of the g-HCH compound lindane and match similar ratios found in air from Alert on Ellesmere Island (Patton et al. 1989, 1991) and snow off Axel Heiberg Island (Hargrave et al. 1988).

Toxaphene was by far the most abundant OC detected in Ellesmere Island lichens, followed by S-PCB, S-HCH, S-CHLOR, S-DDT and S-CBz. Although the most common OC in arctic air and snow is HCH, toxaphene is usually the predominant OC incorporated into arctic animals followed by PCBs (Norstrom et al. 1988; Muuir et al. 1988, 1990; Bidleman et al. 1989). The present data for arctic lichens therefore support these previous animal findings.

The PCB isomers detected in Ellesmere Island lichens were dominated by low chlorinated congeners such as found in air samples from Alert at the northern tip of the island (Patton et al. 1991). This is in contrast to lichens collected from southern Canada which were found to be dominated by the tri- and tetra- PCB congeners (Muir et al. 1993). An increased relative prevalence of high volatility (i.e. low chlorinated) congeners in polar regions is to be expected due to low temperature fractionation and condensation (Wania and Mackay 1993). The cyclodienes heptachlor epoxide, trans- and cis-chlordane, endosulfan and dieldrin were found in the same low abundances in Ellesmere Island lichens as measured in air samples collected throughout the Canadian Arctic (Barrie t al. 1992). The enrichment of the cis- compared to trans-chlordane isomer in Ellesmere Island lichens is in agreement with air samples from the High Arctic (Hoff and Chan 1986). The most common chlordane isomers in arctic lichens (cis-chlordane, trans-nonachlor, cis-nonachlor and heptachlor epoxide) are also those most commonly detected in arctic seals, fish and polar bears (Muir et al. 1990).

Arctic terrestrial food chains are short and direct, leading to the bioaccumulation of trace contaminants. Radionuclides, for example, have been shown to move rapidly through lichen-caribou-predator (human or wolf) food chains in the lower Arctic (e.g. Bird 1968; Hanson 1967). Because of the general agreement between radionuclide and OC abundance within foodwebs (e.g. Kidd et al. 1993), there is a strong likelihood, therefore, that the OC contaminants found within lichens are also present at detectable levels in terrestrial fauna residing in the High Arctic.

Table 2.1. Organochlorine concentrations (mean ± SD, ng/g dry wt) in lichens (Umbilicaria muchlenbergi) *from two locations on Ellesmere Island (76 and 77 °N) in the Canadian High Arctic.*

ELEMENT	**SAMPLING LOCATION**	
	Makinson Inlet	**King Edward Point**
S-CBz	1.0 ±0.2	0.7 ± 0.2
S-HCH	2.1 ± 0.2	1.9 ± 0.4
S-CHLOR	1.5 ± 0.3	1.5 ± 0.5
S-DDT	1.1 ± 0.5	1.1 ± 0.1
S-PCB	4.1 ± 1.7	3.2 ± 1.3
Toxaphene	16.7 ± 2.9	23.7 ± 15.1
Dieldrin	0.7 ± 0.3	0.6 ± 0.1
Endrin	0.1 ± 0.01	0.1 ± 0.01
Mirex	ND	ND
Endosulfan	0.5 ± 0.01	0.5 ± 0.01
PCA	0.1 ± 0.03	0.1 ± 0.01
MECOL	0.1 ± 0.1	0.4 ± 0.5
S-VER	0.1 ± 0.01	0.1 ± 0.01
TRIFLU	0.3 ± 0.5	0.3 ± 0.3

Plate 4. Lichen "rock tripe" from southern Ellesmere Island. Note author conducting sampling in bottom left photo.

CHAPTER 3

Regional Patterns in Organochlorine Contamination of Saxifrage from Ellesmere Island in the High Arctic (77 – 81°N)*

***Adapted from *Environ. Sci. Tech.* 31:1879-1882; Abstract in Appendix II; Coauthors are D.G. Muir and M.D. Segstro**

INTRODUCTION

Determination of the source vectors and bioaccumulation of anthropogenic chemicals in the North American Arctic has received considerable attention (e.g. Barrie et al. 1992; Fellin et al. 1996; Muir et al. 1996). Ellesmere Island, the ninth largest island in the world and the northernmost land mass in North America (76 – 82° N), seems to be a receiving nexus for the deposition of trace contaminants that originate from different long-range transport pathways. Transects of Cesium-137 activity in both soils and plants on Ellesmere Island indicate a south-to-north transport of pollution from mid-latitudinal sources (Hutchison-Benson et al. 1985; Chapter 1). Analysis of lead concentrations and stable isotope ratios in plants and air from Ellesmere Island indicate, in contrast, a north-to-south transport of trans-polar pollution from Eurasian sources (Chapter 6).

Atmospheric and precipitation monitoring of organochlorine (OC) contaminants in the Canadian Arctic has suggested both North American (Hoff and Chan 1986) and European (Welch et al. 1991) sources. Because OCs have been detected in the air, snow, and biota from only localized sampling sites on Ellesmere Island (Gregor and Gummer 1989; Patton et al. 1989, 1991; McNeely and Gummer 1984; Chapter 2), the present study was designed to undertake the first latitudinal transect of OC contamination in the High Arctic to determine the most likely influx pathway of OC pollution to this region.

MATERIALS AND METHODS

Saxifrage (*Saxifraga oppositofolia*) samples were collected from 7 sites located along a transect from 81 to 77° N during spring 1990 as part of the Arctic Light Trans-Ellesmere Island Ski Expedition (Dunn 1991). Additional saxifrage samples (unreplicated or duplicated) were collected from a site in Makinson Inlet on southern Ellesmere Island (77° N) and from another site on nearby Axel Heiberg Island (79°30' N, 87°30' W). Plants were obtained from windswept ridges thought to be free of snow for much of the year. Triplicate samples of nonflowering clumps (both live and dead material) were gathered from areas ranging between 10 and 50 m^2 at each site and stored frozen in aluminum foil envelopes placed within sealed plastic bags. The sealed samples were shipped to the analytical laboratory where they were stored at -40° C until analysis.

In brief, for OC analysis, samples were partially thawed within their sealed bags, and 10-g (fresh wt) amounts were mixed with acetone and held at -10°C for 16 hr to dry the sample and partially extract the OCs. Internal recovery surrogates of aldrin and octachloronapthalene (OCN) were added at the extraction step. Moisture was determined on a separate subsample by drying to constant weight. The dried saxifrage residue was then blended with dichloromethane. Extracts were evaporated, exchanged into hexane, and fractioned on Florisil columns. Organochlorines were measured by capillary gas chromatography using procedures described in Muir et al. (1993, 1995) and Chapter 2. Toxaphene and other OCs were confirmed by high-resolution electron ionization mass spectrophometry (HREIMS) or electron capture negative chemical ionization MS (Chapter 2). Total PCBs (ΣPCB) consisting of 85 peaks (including co-eluting congeners) were quantified using external standards (NRC, Halifax, NS; Ultra Scientific Hope, RI) as described in Muir et al. (1993, 1995) and Chapter 2. Cesium-137 activity was measured on a GE counter as described in Chapter 1.

Recovery efficiencies of the internal standards aldrin and OCN ranged from 85 to 95%. Blank analyses were free of major interfering GC-ECD peaks, which permitted low detection limits (Chapter 2). Data presented here are for only the most abundant OC residues found in plants from Ellesmere Island: toxaphene, ΣHCH (sum of α-, β-,and γ-hexachlorocyclohexane isomers),ΣCBz (sum of 1,2,4,5-tetrachlorobenzene (T_4CBz), 1,2,3,4-T_4CBz, petachlorobenzene, and hexachlorobenzene (HCBz), ΣCHL (sum of chlordane-related compounds *cis*- and *trans*-chlorodane, *cis*- and *trans*-nonoachlor, heptachlor, heptachlor epoxide, nonachlor III, and C5), ΣDDT (sum of *o,p*'-DDE, -DDD, and –DDT), and selected PCB homologs.

Results and Discussion

With the exception of ΣPCBs (not shown), the other OCs displayed decreases in saxifrage accumulation with northward distance on Ellesmere Island (Figure 3.1). In this respect, patterns in OC accumulation in saxifrage on Ellesmere Island are similar to that shown by ^{137}Cs (Chapter 1) but not lead (Chapter 6). This evidence suggests that the primary source of OCs to this region of the High Arctic is through atmospheric deposition from mid-latitudinal North American sources and not as a result of trans-polar deposition from Eurasian sources.

Other studies have recorded northward decreases in OC residues in northern hemisphere biota (e.g. Norstrom et al. 1988; Muir et al. 1990; Addison and Zinck 1986) and sediments (Muir et al. 1995, 1996) and have interpreted these latitudinal gradients as being due to temperature-related processes of global fractionation and condensation (Calamari et al.

1991; Wania and Mackay 1993). In the present case, we believe that the limited 4° latitudinal range of our sampling transect is too narrow for differential fractionation processes to account for the observed pattern in OC accumulation shown in Figure 3.1. Instead, we suggest that differences in the saxifrage accumulation of OCs with latitude merely reflect consistent regional patterns in the deposition of southerly-originating pollutants.

Expressing OC accumulations with respect to ^{137}Cs activity for saxifrage samples allows for inter-site comparisons to be made that accommodate differences in the localized deposition of trace contaminants due to variable precipitation. In the present case, this resulted in the reordering of two sampling sites that had been formerly based on latitude in Figure 3.1. Organochlorine concentrations in Ellesmere Island saxifrage were found to be related to plant ^{137}Cs activities, although steeper slopes existed for ΣDDT and toxaphane than for HCH, ΣCBz, ΣCHL (Figure 3.2), and ΣPCB (Figure 3.3). Within the PCB group, octa- and heptachlorobiphenyls (not shown) were found to have steeper slopes in relation to ^{137}Cs than hexa-, penta-, tetra-, and trichlorobiphenyls (not shown). Other studies that have used actual precipitation data to estimate plant uptake of OCs have met with only limited success (Muir et al. 1993; Thomas 1984).

Independent samples from Makinson Inlet on Ellesmere Island (77° N) and across from Eureka at Buchanan Lake on Axel Heiberg Island (79° N) indicate that not only were all OC residues lower at the more northern site in support of the transect data in Figure 3.1 but also that the relative differences between the two sites were greater for toxaphene and DDT than for any of the other OCs as would be expected from the trends shown in Figures 3.2 and 3.3 (Figure 3.4).

Inter-site variability in OC accumulation in saxifrage was not related to differences in either the physical characteristics of the sampling sites (distance from the coast, elevation, substrate typology, site topography, immediate surface moisture conditions, windbreaks or obstructions within 100 m, surrounding area characteristics within 1000 m, and site orientation) or to differences in plant growth efficiencies due to water content as measured by stable carbon isotope analysis (R. France, unpubl. data).

We presently have no detailed explanation as to why ^{137}Cs activity is more closely associated with toxaphene, ΣDDT, and the higher chlorinated PCB homologs than it is with ΣCBz, ΣCHL, and the lower chlorinated PCB homologs in saxifrage, although it is possible that differing partitioning of these elements to the plant surfaces may be responsible. Saxifrage, a vascular plant, has a waxy outer surface on its branches making it ideal for scavenging airborne organics. The correspondence of high ^{137}Cs with toxaphene and ΣDDT implies that these less volatile OC, which have high proportions in the particle phase during winter temperatures, may reach the plants via a common (particle deposition) pathway. Although high volatility OCs (i.e. HCH and HCB) predominate in polar air (Fellin et al. 1996; Patton et al. 1991), they also readily volatize from plant and soil surfaces because of their high vapor pressures and lower plant-air partition coefficients. It is the less volatile, more chlorinated components that are retained on the plants in disproportionate amounts relative to their concentrations observed in air at Alert on northern Ellesmere Island (Fellin et al. 1996; Patton et al. 1991). Lead et al. (1996) found that over time (1977-1990) the proportion of higher chlorinated PCBs increased in moss samples form northern Norway relative to sites in the south.

The capability of environmental scientists for measuring ^{137}Cs accumulation is much easier than that for accurately quantifying OC contamination. The close association between residues of these two micropollutants in saxifrage from Ellesmere Island, despite probable differences in their precise atmospheric transport processes and vegetation partitioning mechanisms, raises the possibility that ^{137}Cs uptake could be used as a surrogate environmental monitor for studies on regional patterns in OC deposition in those locations where the overall source pathways are similar. Although there is a general agreement between both ^{137}Cs and OC abundance within components of aquatic ecosystems (Rowman and Rasmussen 1994), difficulties in developing empirical conversions between the two is dependent on understanding the influences of water chemistry and metabolic functions (Rowman and Rasmussen 1995) and how these are affected by foodweb interactions (Muir et al. 1996; Kidd et al. 1995). Sampling terrestrial plants (Calamari et al. 1991; Simonich and Hities 1995) and, even better, surface soils or lake sediments (Muir et al. 1995, 1996), circumvents foodweb difficulties and may offer promise toward increasing our understanding of geographic trends in OC deposition. As data already exist on latitudinal patterns in the atmospheric concentration (Hardy et al. 1968; Feely et al. 1978), terrestrial deposition (Hutchison-Benson et al. 1985), plant uptake (Hutchison-Benson et al. 1985; Chapter 1), and biogeochemical processing (Taylor et al. 1985) of radionuclides, development of empirical relationships between these measurements with similar ones for OCs would be a profitable direction worth pursuing further.

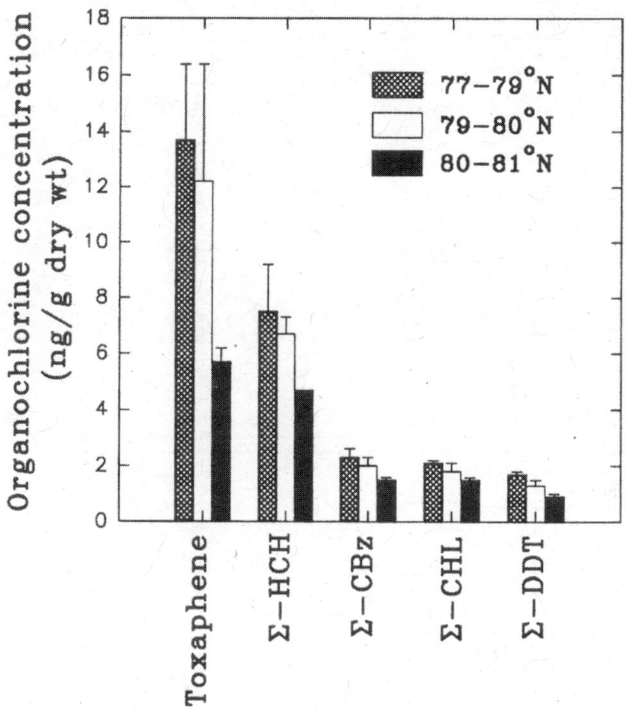

Figure 3.1. Variability in organochlorine contamination of saxifrage in relation to latitude on Ellesmere Island. Means + SEs are based on two, three, and two independent sample sites for respective south-to-north latitudinal categories. Organochlorine contamination in plants at 80–81° N was significantly different ($p < 0.05$, Duncan's multiple range test, ANOVA) from those at 79–80° N or 77–79° N, which in turn did not differ.

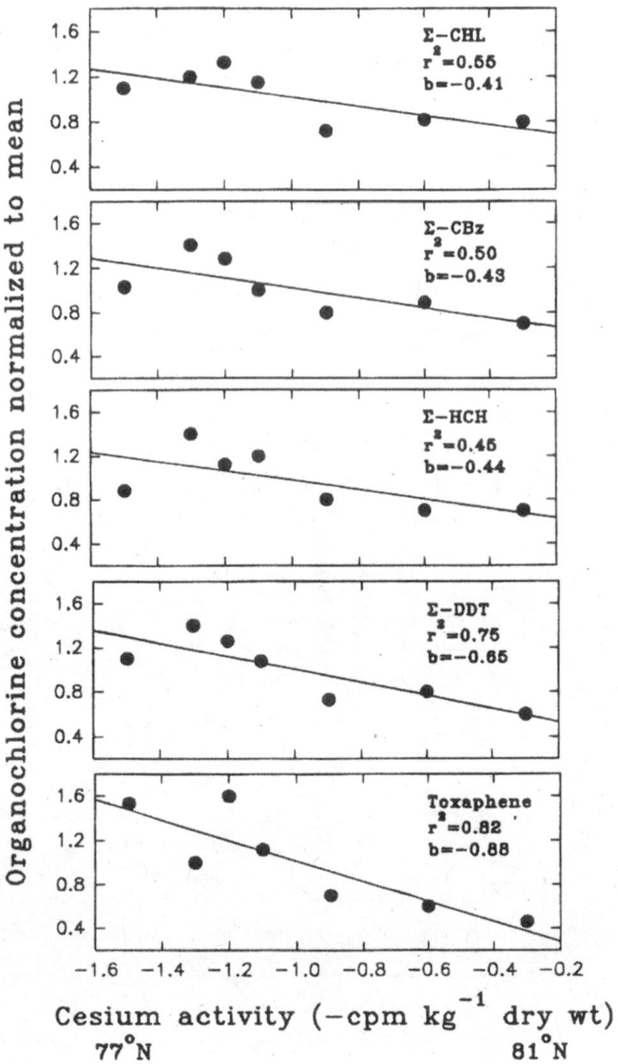

Figure 3.2. Relationships between organochlorine concentrations (mean normalized to facilitate comparison among different absolute concentrations) and cesium-137 activities in samples of Ellesmere Island saxifrage. All slopes are significant at $p = 0.01$. Presented OC data are site-specific means, with average per site SDs (not shown for clarity) of ± 13% of these means.

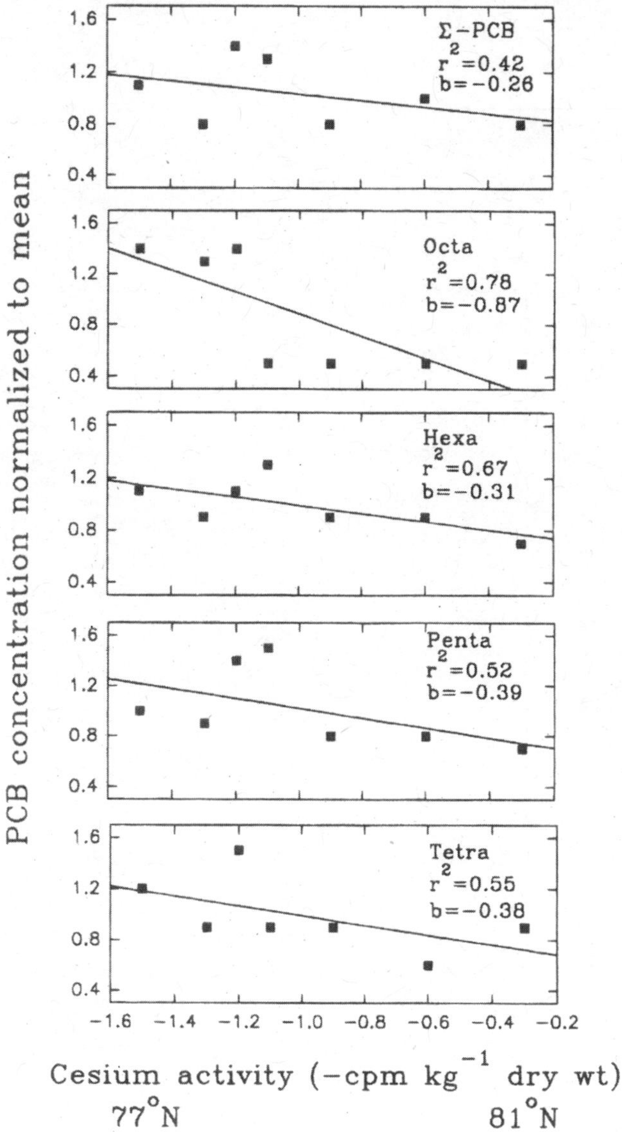

Figure 3.3. Relationships between slopes of PCB and selected PCB homologs (mean normalized to facilitate comparison among different absolute concentrations) and cesium-137 activites in samples of Ellesmere Island saxifrage. Slopes for octa- and penta-chlorobiphenyls are significant at $p = 0.05$. Presented data are site-specific means, with average per site SDs (not shown for clarity) of $\pm$ 16% of these means.

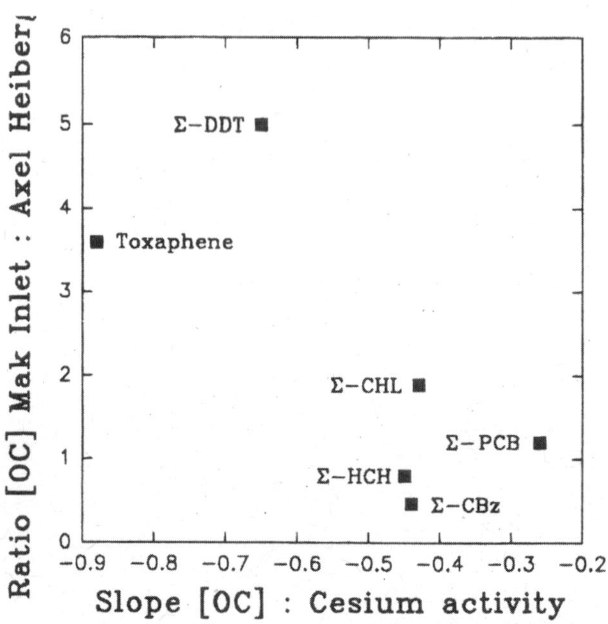

Figure 3.4. Scatter plot between slopes of organochlorine concentrations to cesium-137 activities from Figures 3.2 and 3.3, and ratios between OC concentrations in saxifrage from a site in Makinson Inlet on Ellesmere Island (77° N) compared to a site on Axel Heiberg Island (79° N). Organochlorines having steeper slopes with respect to latitude (toxaphene and DDT) display greter differences between these two additional sampling sites than do the other OCs with shallower slopes. Ratios for individual PCB homolog groups (not shown) ranged from 0.8 to 1.3.

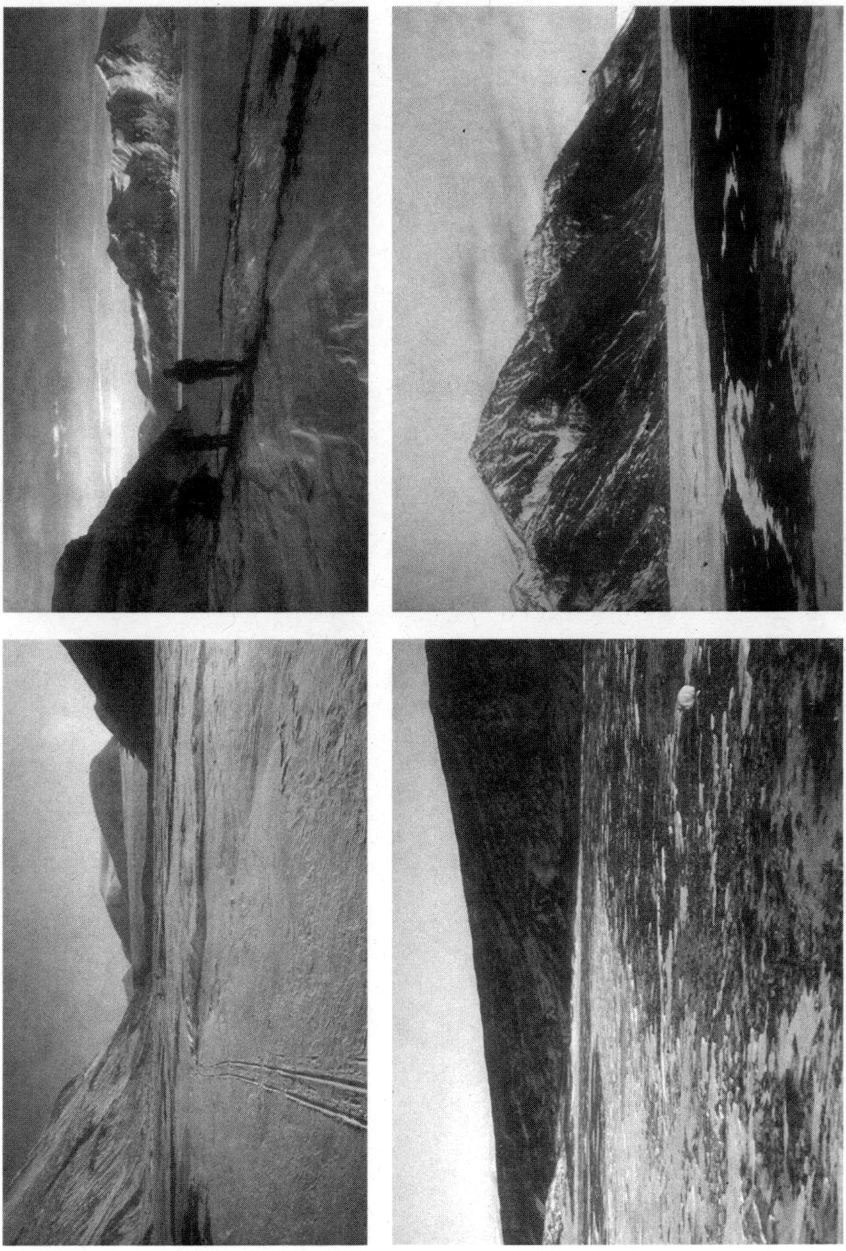

Plate 5. Sample sites from northern Ellesmere Island: Dodge Valley (bottom-left), Archer Fiord (top-left), and Radmore Harbour (top-right, bottom-right).

CHAPTER 4

Elevated PCB Contamination of Coastal Plants Near Polynyas in the High Arctic*

***Adapted from *Bull. Environ. Contam. Toxicol.* 59:76-82.**

INTRODUCTION

During the Trans-Ellesmere Island Ski Expedition, coastal samples of saxifrage (*Saxifraga oppositofolia*) were collected from 81 to 77°N to measure the long-range transport and deposition of radiocesium (Chapter 1), lead (Chapter 6), and organochlorines (Chapter 3) to this region of the High Arctic. Preliminary analysis of sporadically collected lichens had revealed the most abundant organochlorine (OC) contaminants on Ellesmere Island to be toxaphene, Σ-PCBs (sum of polychlorinated biphenyls), Σ-HCHs (sum of hexachlorocyclohexanes), Σ-CHLOR (sum of chlodanes), Σ-DDT (sum of DDT isomers), and Σ-CBz (sum of hexachlorobenzenes) (Chapter 2). Saxifrage transect data showed that although all these OCs displayed northward decreases in their concentrations which matched those of radio-cesium in the plants, the relationship for Σ-PCB had the lowest slope and largest amount of variance left unexplained (Chapter 3). Radiocesium activity was itself thought to be a convenient estimator of the cumulative precipitation and deposition of southerly originating trace contaminants.

Several studies have likewise noted dissimilar regional patterns in PCB deposition and uptake compared to those simultaneously determined for other OCs (e.g. Carlberg et al. 1983; Norstrom et al. 1988; Larsson and Okla 1989; Larsson et al. 1990; Muir et al. 1993). The purpose of the present study was to investigate the somewhat divergent pattern of PCB uptake by saxifrage on Ellesmere Island. These represent some of the first data concerning PCB concentrations in vascular plants from anywhere in the High Arctic.

Second to the latitudinal range, the most obvious difference that existed among sampling sites was whether or not they were situated beside polynyas. Polynyas are annually recurring areas of open water surrounded by ice that are kept from freezing by the combined effects of wind, tidal fluctuations, and upwellings (Sterling and Cleator 1981), and are of critical importance in providing winter refugia for marine mammals and as spring staging grounds for migrant birds (e.g. Sterling and Cleator 1981; Chapter 8).

Proximal association with polynyas was thought to be of putative importance to plant bioaccumulation due to the high water-air transfer volatilization of PCBs (e.g. Anderson and Parker 1990; Achman et al. 1993) which has been suggested to exacerbate the deposition from long-range sources at coastal compared to inland sites in Sweden through the localized effects of sea-spray aerosols (Larsson and Okla 1989). My expectation was therefore for a similar pattern to exist in the High Arctic such that saxifrage located near polynyas would exhibit higher tissue residues of PCBs than those from plants collected elsewhere.

Materials and Methods

Saxifrage samples were obtained from 7 coastal sites along a transect from 81 to 77°N. Particulars concerning site selection, location, and sample collection are described in Chapter 1. Two of the sampling sites were situated nearby the small polynyas of Flagler Bay and Makinson Inlet (Chapter 8). In addition to latitude and presence of polynyas, the following information was documented for each sampling site (Chapter 1): distance from coast, elevation, substrate typology, site topography, plant water content, windbreaks or obstructions within 100m, surrounding area characteristics (1000m), and site orientation. Triplicate samples were analyzed for PCBs using procedures described in Muir et al. (1993).

Results and Discussion

The average Σ-PCB concentration of saxifrage from the two polynya sites was found to be significantly higher (paired t-test, $p < 0.05$) than that of plants from the 5 other non-polynya sites (Figure 4.1). When Σ-PCB residues are expressed in relation to radiocesium activity from Chapter 1 to standardize for inter-site differences in the long-range northward transport of trace pollutants, the increased PCB contamination of saxifrage from polynya sites remained evident. Elevated Σ-PCB concentrations for saxifrage from polynya compared to non-polynya sites were due to higher abundances of all measured homologs. The only physical site variable that was found to be significantly different between the two data groupings was the water content of plant samples: $10.6 \pm 1.4\%$ for poynya sites and $7.9 \pm 1.0\%$ for non-polynya sites.

The higher Σ-PCB concentration in saxifrage located nearby permanently open bodies of water on Ellesmere Island is in agreement with the expectations predicted from Larsson and Okla's (1989) work on fallout in Sweden.

There is no given, however, that elevated toxicant concentrations will always exist for plants near coastal water. For example, it is possible that localized meteorological differences may supercede any effect of volatilization from surface waters. Data from Larsson and Okla (1989) indicate that site-specific differences in the amount of precipitation can produce differences in PCB fallout of an even greater magnitude than that believed to be produced through coastal sea-spray effects. These authors supposed that such locally produced clean precipitation could wash-out any long-range effects and thereby obscure demonstration of systematic regional patterns in PCB accumulation.

For Ellesmere Island, Gregor and Gummer (1989) determined that the coastal oasis of Alexandra Fiord, located nearby the Flagler Bay Polynya, was the cleanest of 12 arctic sites with respect to snow PCB concentrations. The loosely packed snow found at Alexandra Fiord implied that the area was receiving larger quantities of locally induced precipitation than the other sites. This was supposed by Gregor and Gummer (1989) to result in a dilution of the concentration of pollutants transported to the arctic from long-range sources so that the PCB content of the snow at Alexandra Fiord was only 2% of that found at the Aggasiz Ice Cap located 175 km further north on Ellesmere Island. Other coastal sites in the Canadian Arctic were also considered to be less polluted compared to inland or snowbound sites.

It would be erroneous, however, to simply regard polynyas as being typical arctic coastal sites. Polynyas, through the act of maintaining open water at just above freezing despite being surrounded by substantially colder winter air, create an enormous thermal vapour flux which might be expected to produce a localized microclimate very dissimilar from that for coastal regions that are ice-bound for up to ten months of the year (Sterling and Cleator 1981). Condensation of this vapour in the form of snowfall might explain why saxifrage collected from near the two polynyas in the present study were significantly wetter than plants obtained from other areas.

Although it had been the intention of the present sampling design to collect only plants growing on exposed windswept ridges or beaches that were free of snow for much of the year, this was easier to accomplish in early May at the start of sampling when snow was abundant than it was in late June toward the end the Expedition when snowmelt was proceeding and therefore purposeful selection of year-round, snow-free sites was more problematic. The increased water moisture of plants from the polynya sites may imply a greater supply of snow cover to such locations. This is important because accumulation of radiocesium has been shown to be reduced for arctic vegetation collected from such mesic sites (Hutchison-Benson et al. 1985) presumably as a result of the protective cover of snow. Because PCB uptake by plants proceeds through foliar assimilation due to atmospheric transfer rather than by root relocation, it may be inferred that the higher snow possibly associated with polynyas could likewise produce lower toxicant residues in plants growing there. At the same time, because much of the High Arctic can be regarded as a desert, any increased moisture is very beneficial to plant growth and system productivity. Despite the possibility of a resultant growth dilution of saxifrage near polynyas due to increased moisture present in the area, the observation that Σ-PCB concentrations were still higher than in inland sites implies an overwhelming influence of increased volatilization at polynyas.

A higher percentage of PCBs in seawater are absorbed to suspended particles at high latitudes (Tanabe et al. 1983). Generally low concentrations of suspended particulate matter in the arctic water column leads to low rates of sedimentation and therefore increased residence times of particulate-reactive chemicals. As a result, there is normally little chemical transformation that occurs when PCBs and other OCs are transferred between air, snow, and seawater compartments (Hargrave et al. 1988). In contrast, in situations of high productivity such as may occur under fast-ice, it is the higher chlorinated PCBs that will be preferentially removed from the polar water column due to increased sedimentation of particles (Tanabe et al. 1983). This may explain why plants located near open water polynyas have elevated concentrations of high chlorinated homologs compared to plants located near ice-bound shorelines where sedimentation and removal of these homologs might be higher. The dichotomy between the two locations in terms of low cholrinated homologs is still, however, unexplained by this speculative scenario.

In the end, we are left with some as yet unknown factor such as perhaps a difference in the rate of volatilization between polynya and non-polynya sites which might explain the difference in PCBs accumulated by respective proximal terrestrial plants. Although volatilization is dependent on water temperature and therefore might be expected to be lower or absent in the Arctic (Iwata et al. 1993), the rapid wind movements often associated with keeping the water of polynyas free of ice may serve to maintain a role of these areas as regional sources rather than as sinks for PCBs. Given the much longer time period of water-air volatilization which can take place at polynyas compared to other, typically ice-bound arctic coastlines, the increase in low chlorinated, high volatility PCB homologs in saxifrage from near polynyas might simply be a result of the presence of such PCB homologs in the openwater followed by their ongoing removal and overland dispersal.

Obviously, the previous hypotheses will remain only speculations until detailed sampling of PCB transfers is undertaken in polynyas. Given the acknowledged importance of such regions to sustaining productivity and biodiversity in the High Arctic (e.g. Sterling and Cleator 1981; Chapter 8), direction of environmental sampling effort to polynyas would be a profitable future direction worth pursuing in a much more systematic fashion than the present effort.

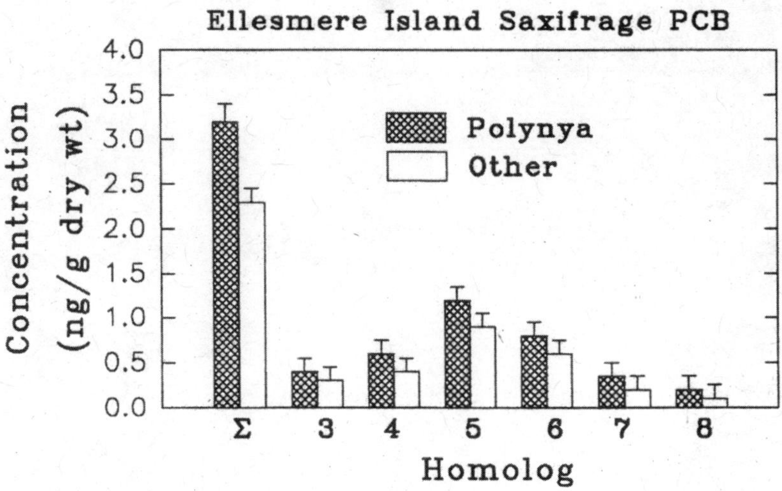

Figure 4.1. Concentrations of summed PCBs and individual homologs (ng/g dry wt) of saxifrage samples collected from polynya and other sites on Ellesmere Island (data are means ± SEs).

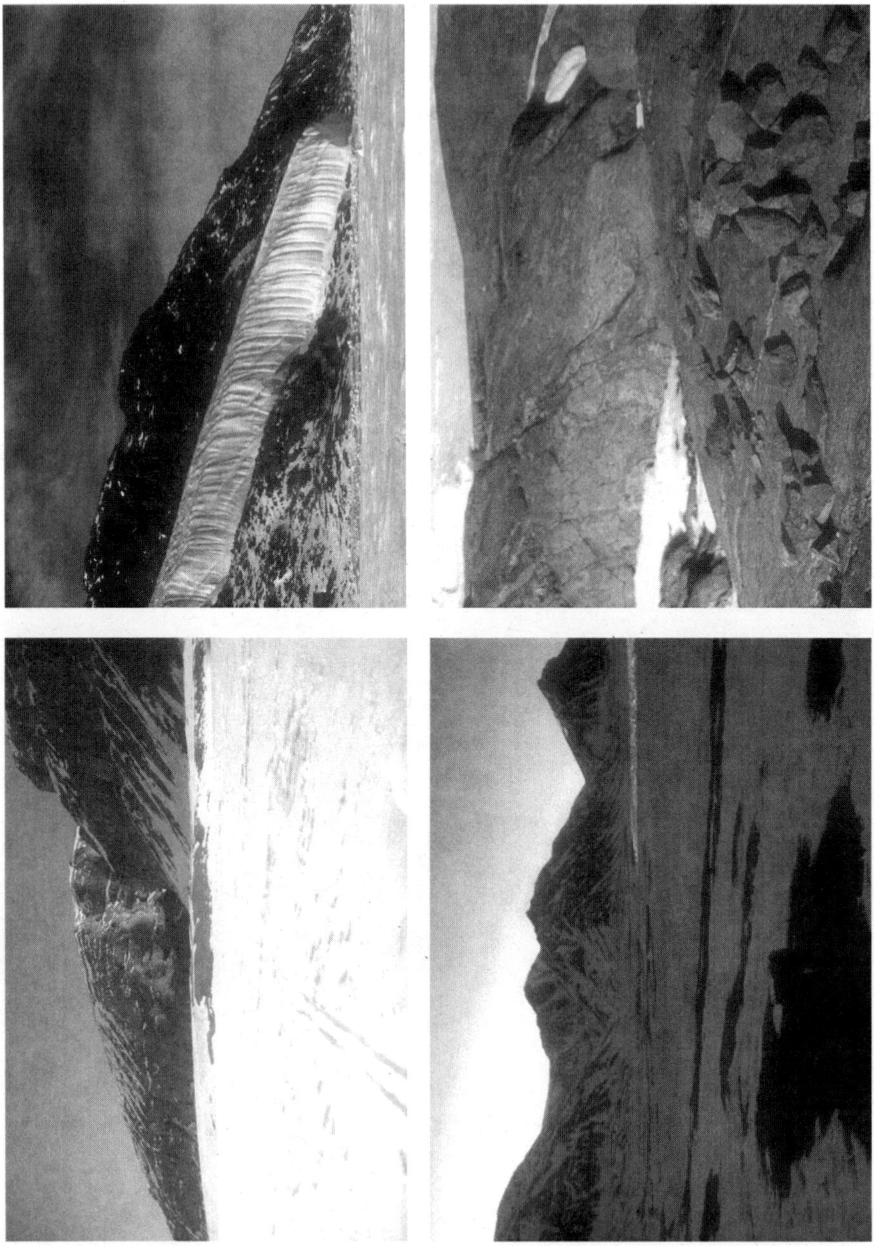

Plate 6. Sample sites from central Ellesmere Island: Joiner Bay (bottom-left), Norman Lockyer Island (top-left), Flagler Bay Polynya (bottom-right) and Jokel Fiord (top-right). Note *Paleoeskimo ruin in bottom-right photo.*

CHAPTER 5

Lead Concentrations in Lichens from the Canadian High Arctic in Relation to the Latitudinal Pollution Gradient*

***Adapted from *Wat. Air Soil Pollut.* 90:469-474; Abstract in Appendix II; Coauthor is M. Coquery**

INTRODUCTION

The expression "pristine site" has become anachronistic. Nowhere is free of wither the overt (Benton 1991: Chapter 14) or insidious (Nriagu and Pacyna 1988; Focardi et al. 1991) by-products of human society. Lead concentrations have, for example, increased markedly in the Arctic over the last century (Murozumi et al. 1969). Determination of the transport pathways of anthropogenic chemicals is, as a result, a pressing concern of environmental scientists working in the Arctic (e.g. Rahn 1981a,b; Lowenthal and rahn 1985; Sturges and Barrie 1989: Sturges et al. 1993).

North American latitudinal gradients in bioaccumulation have proven useful in identifying a south-to-north source vector for several contaminants (e.g. Wania and Mackay 1993: Hutchinson-Benson et al. 1980) but presently have not been established for lead (Pb). In Scandinavia, however, Pb concentrations in mosses and lake sediments decline along northward transects (Ruhling and Tyler 1973; Pakarinen and Tolonen 1976; Johansson 1989), and can be directly related to parallel latitudinal trends in contamination of wet deposition (Ross 1987). Canadian data, in contrast, are more limited but do suggest a pattern of lower concentrations to the north. Sediment Pb contents were determined to be highest in south-central Ontario lakes, intermediate in southern Quebec lakes, and lowest in northern Quebec lakes (Oulett and Jones 1983; Evans and Dillon 1985). Lead concentrations were lower in *Sphagnum* from two sites in south-central Northwest Territories than from either a site in central Alberta (Pakarinen and Tolonen 1976) or one in northern Ontario (Glooschenko and Capabianco 1978). Lead concentrations in moss collected from a southern Quebec site were higher than those for two locations in central Quebec, which in turn were higher than those for two other sites in northern Quebec (Groet 1976).

Although determinations of Pb accumulation in lichens from southern Canada are frequent (e.g. Zakshek et al. 1986), such data are extremely rate at higher latitudes with, for example, 5 collections (east of 100° W) from 60 to 70° N, a further 5 sites from 70 to 75° N, and only three sites in the High Arctic north of 75° N (Puckett and Finegan 1980).

The purpose of the present study was: (a) to determine the Pb content of 5 lichen species from two locations on Ellesmere Island in the High Arctic, and then (b) to compare these data with those collected from previous studies toward defining the latitudinal gradient of Pb concentration in lichens throughout Canada.

Methods

Lichens were collected in July 1990 from two sites on Ellesmere Island in the Canadian High Arctic: 4 species from Sverdrup Pass in the central portion of the island at about 79° N, and 5 replicates of one species from King Edward Point on the southernmost tip of the island at about 76° N. Particulars concerning site selection and location, and sample collection are described in Chapter 1.

To minimize metal contamination, all plastic and glassware were washed thoroughly, soaked in 15% HNO_3 for 24 hr, and then rinsed 7 times with distilled deionized water. Lichen samples were cleaned with forceps to remove material adhering to their surfaces and dried at 60° C for 24 hr. All dried lichen samples were ground in a Wiley mill (0.5 mm mesh) and stored at room temperature. Prior to analysis, lichen samples were redried for 12 hr at 60° C and placed in a desiccator. About 200 mg of lichen material was weighed and digested with 2.5 ml of HNO_3 in Pyrex boiling tubes. The tubes were covered with glass bulbs and left for 12 hr at room temperature. After cooling, the digests were diluted to 25 ml with distilled deionized water (final concentration 10% HNO_3). Solutions were analyzed for Pb by flameless atomic absorption spectrophotometry with a Varian SpectrAA-30 and graphite furnace Varian GTA-96. Reagent blanks were carried through all procedures. The accuracy of the methods was verified by subjecting standard reference material (U.S. National Bureau of Standards (NBS): Citrus leaves 1572 and TORT) to the overall analytical procedures.

Results and Discussion

Quality assurance of our analytical procedures was high. Procedure blanks averaged 1 $\mu g\ L^{-1}$ which was subtracted from all lichen results. Citrus values were 13.5 ± 0.3 (SD) $\mu g\ g^{-1}$, and when ground like the lichen samples, were 12.6 ± 1.5 $\mu g\ g^{-1}$. Both of these values are within the NBS certified range of 13.3 ± 2.4 $\mu g\ g^{-1}$. Likewise, a single TORT sample yielded 12.4 $\mu g\ g^{-1}$ within the NBS certified range of 10.4 ± 2.0 $\mu g\ g^{-1}$.

Lead concentrations in the four lichen species collected from Sverdrup Pass ranged from 1.2 to 5.0 μg Pb g^{-1} with a mean of 2.9 + 1.4 (SD) (Table 5.1). The 5 samples of *Umbilicaria muhlenbergii* from King Edward Point ranged from 1.3 to 3.5 μg Pb g^{-1} with a mean of 2.0 ± 0.8 (SD). All these values are very low and, with the exception of two Sphagnum samples obtained from a single site southwest of Great Slave Lake, N.W.T. (Pakarinen and Tolonen 1976), are the lowest Pb concentrations that have ever been recorded in lichens or mosses from high latitude sites (Table 5.2).

When the results from previous lichen biomonitoring studies for eastern Canada (all of which used X-ray fluorescence) are combined, and expressed as means for 5° bands of latitude, the overall distribution pattern of Pb deposition can be established (Figure 5.1). Lead in lichens, like Cesium-137 in soil (Hutchison-Benson et al. 1980), displays a progressive decline in concentration with increasing latitude in Canada. Similar results have also been clearly demonstrated for Pb in mosses within Scandinavia, and suggested by the few data for Pb in Canadian mosses (Figure 5.1), but see Chapter 6 for regional differences with respect to this general relationship.

The present data on Pb accumulation in lichens are lower than those determined previously by Puckett (1978) for the same latitudes (Figure 5.1). At least two possible factors could contribute toward this discrepancy. Decreased use of leaded gasoline in North America over the interlying years has reduced the import of Pb to certain regions of the Arctic (Boutron et al. 1991). Also, the elevated concentrations in the previous study may have been due to their determination through X-ray fluorescence (without quality assurance with respect to standards) rather than atomic absorption spectrophotometry (and corresponding quality checks) as used in the present study. We therefore consider that our overall average of 2 $\mu g\ g^{-1}$ (close to the analytical detection limit of 1 $\mu g\ g^{-1}$) represents the present-day uncontaminated background concentration of Pb in lichens from these remote High Arctic locations.

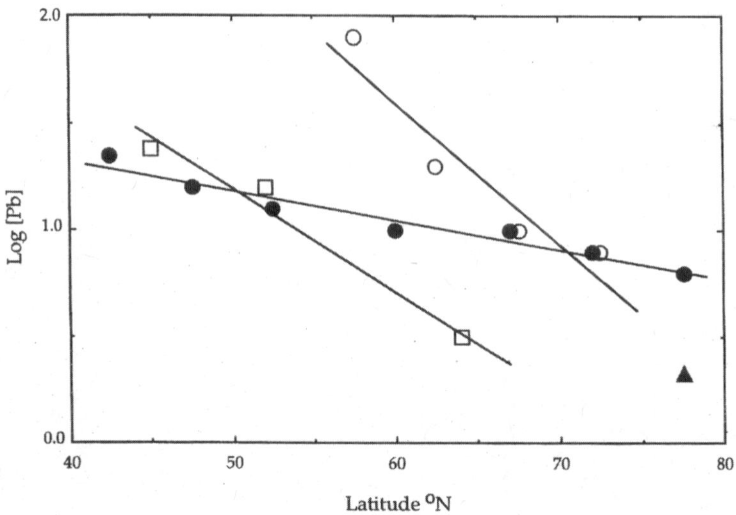

Figure 5.1. Latitudinal distribution of lead deposition in eastern Canada as measured by lichen (closed circles) and moss (open squares) accumulation, and in Scandinavia as measured by moss (open circles) accumulation. Data sources for Canadian lichens (all determined by X-ray fluorescence) are Puckett (1978), Puckett and Finegan (1977), Niebor et al. (1972) and Zakshek et al. (1986); for Canadian mosses (as determined by atomic absorption spectrophotometry) are Pakarinen and Tolonen (1976), Groet (1976), Glooschenko and Capabianco (1978) and Folkeson (1979); and for Scandinavian mosses (as determined by atomic absorption spectrophotometry) are Ruhling and Tyler (1973), Pakarinen and Tolonen (1976) and Ross (1987). Data from the present study for lichens (as determined by atomic absorption spectrophotometry) are denoted by the closed triangle.

Table 5.1. Lead concentrations (µg g^{-1} dry wt) in lichens from Ellesmere Island, N.W.T. as measured by flameless atomic absorption spectrophotometry.

Location	Species	[Pb]
Sverdrup Pass	*Umbilicaria deusta*	4.0
	Umbilicaria sp.	5.0
	Alectoria ochroleuca	2.3
	Cetraria nivalis	1.2
	Cetraria nivalis	2.3
King Edward Point	*Umbilicaria muhlenbergii*	1.6
	Umbilicaria muhlenbergii	2.1
	Umbilicaria muhlenbergii	3.5
	Umbilicaria muhlenbergii	1.4
	Umbilicaria muhlenbergii	1.3

Table 5.2 : Comparison of average lead concentrations (µg g^{-1} dry wt) in moss (M) and lichen (L) collected from remote locations in northern Scandinavia and the Canadian Arctic. All analyses by atomic absorption spectrophotometry except Puckett and Finegan (1980) and Zashek et al. (1986) which are by X-ray flourescence.

[Pb]	Plant	Location	References
11	M	Northern Scandinavia	Ruhling and Tyler 1971
10	M	Northern Sweden	Ross 1990
8	M	Northern Scandinavia	Ruhling and Tyler 1973
8	M	Northern Finland	Pakarinen and Tolonen 1977
7	M	South-western N.W.T.	Glooschenko and Capobianco 1978
5	L	Northern Quebec	Zakshek *et al.*, 1986
5	L	N.W.T. overall	Puckett and Finegan 1980
4	M	Northern Finland	Pakarinen and Tolonen 1976
2	L	Ellesmere Island, N.W.T.	This Study
1	M	South-western N.W.T.	Pakarinen and Tolonen 1976

Plate 7. Sample sites from southern Ellesmere Island: Makinson Inlet Head (bottom-left), Jones Sound (top-left), Craig Harbour (bottom-right), and King Edward Point (top-right).

CHAPTER 6

Lead Concentrations and Stable Isotopic Evidence for Transpolar Contamination of Plants in the Canadian High Arctic*

***Adapted from *Ambio* 27:506-508; Abstract in Appendix II; Coauthor is J.M. Blais**

INTRODUCTION

Anthropogenically derived lead, a major constituent of the aerosols colloquially referred to as "arctic haze" (Barrie et al. 1992), is deposited at high latitudes through precipitation (Muir et al. 1996; Boutron et al. 1991). Because the isotopic composition of lead varies among different ore bodies, the relative contributions from different sources can be estimated (Sturges and Barrie 1987). Here we present measurements of $^{206}Pb/^{207}Pb$ ratios and total lead concentrations in pants collected along a transect on Ellesmere Island (76 – 82° N) to provide the first evidence for the incorporation into Canadian biota of contaminants arising from Eurasian sources.

Compilation of literature data on the sulfur content of North American lichens (Figure 6.1) suggests that contamination of terrestrial foodwebs in the High Arctic may be occurring as a result of transpolar pollution. Ellesmere Island, the northernmost landmass in North America, is approximately equidistant from both North American and Eurasian SO_2 emission sources (Figure 6.1), and may therefore be a deposition nexus for receiving artic haze pollutants (Figure 6.2).

Aerosol sampling from north Ellesmere Island has indicated the presence of high lead concentrations and provided evidence, based on stable isotope ratios, that much of this lead originates from Eurasian sources (Sturges and Barrie 1989). Unfortunately, aerosol sampling often involves dealing with considerable temporal variability as a result of meteorological vagaries, sometimes making it difficult to clearly identify dominant sources (Church et al. 1990) due to lack of cumulative integration (Blais 1996; Shirahata et al. 1980). Integrative biomonitoring of lead pollution through isotopic source differentiation has until now been restricted to low-to-mid latitudes (Smith et al. 1990; Rabinowitz and Witherill 1972; Ault et al. 1970).

METHODS

Plants (*Saxifraga oppositofolia*) were collected from 10 locations along a 6° north-to-south transect on Ellesmere Island during the 1990 Arctic Light Trans-Ellesmere Island Ski Expedition (Dunn 1991). Four to five samples of clumps of saxifrage were gathered from areas ranging from 10 and 50 m^2 for each site. Both live and dead material—the latter which can contribute up to 75% of the biomass of saxifrage (Stewart 1984)—were sampled as free of soil as possible from exposed windswept ridges assumed to be devoid of snow for much of the year (Fig. 6.3). Samples were stored in sealed, reinforced bags. There were no differences in estimated growth dilution and, therefore, physiologically-altered bioaccumulation rates, for saxifrage sampled within this latitudinal range (Chapter 1; Taylor et al. 1985).

Lead concentrations were analyzed on a Varian SpecAA-30 and graphite furnance Varian GTA-96 as described in Chapter 4. Lead isotopic measurements were made by ICP-MS as described in Blais (1996). In both cases, all samples were analyzed in conjunction with standard reference materials (NBS, citrus leaves, TORT). Quality assurance of our analytical procedures was high, averaging ± 1.0 ppm dry wt for lead concentrations (Chapter 5) and ± 0.0058 for $^{206}Pb/^{207}Pb$ ratios (Blais 1996). Other lead isotope ratios, though measured, were not considered here because the error terms were larger and results more open to different interpretation.

Results and Discussion

Lead concentrations in saxifrage were found to be threefold higher north, compared to south, of 80° N on Ellesmere Island (Figure 6.4). Expressing lead concentrations with respect to ^{137}Cs activities for the saxifrage samples (Chapter 1) allows for intersite comparisons to be made which accommodate differences in the localized deposition patterns of southerly-originating contaminants (Hutchison- Benson et al. 1985) due to variable precipitation (Chapter 3). This analysis confirms that lead concentrations in saxifrage samples increase toward the north on Ellesmere Island (Figure 6.4). Lead concentrations of $< 2\ \mu g\ g^{-1}$ for saxifrage from south-central Ellesmere Island agree with lichen data collected from the same region at the same time and confirm that the northward influence of Pb pollution from mid-latitudinal sources in eastern North America only weakly extends into the High Arctic (Chapter 5).

Average isotopic rations ($^{206}Pb/^{207}Pb$) in saxifrage decreased from 1.227 between 78 - 79° N, to 1.215 between 79 - 80° N, and to 1.190 between 80 - 82° N (Figure 6.5). Ratios ≥ 1.20 are indicative of United States gasoline sources (Sturges and Barrie 1987, 1989; Hopper and Barrie 1988) and have previously been detected in snow cores from Greenland (Rosman et al. 1993). In contrast, lead isotope ratios for saxifrage from northern Ellesmere Island are below these United States sources and are instead comparable to values documented for aerosols and snow in this region (Sturges and Barrie 1986) which have been ascribed for Ellesmere Island as well as elsewhere (Alaska – Sturges et al. 1993; Svalbard – Sturges and Barrie 1989; Greenland – Rosman et al. 1993) as being partially due to Eurasian sources with values between 1.14 and 1.19 (Barrie et al. 1992; Sturges and Barrie 1989;

Sturges et al. 1993). Confirmed isotopic sources for Canadian gasoline have ratios of about 1.15 (Sturges and Barrie 1987, 1989; Hopper and Barrie 1988), in other words indistinguishable from Eurasian sources. However, due to meteorological conditions as well as limited production, such Canadian sources are not though to be transported to or to contribute toward contamination at high latitudes relative to those originating from the United States (Barrie et al. 1992; Sturges and Barrie 1989; Sturges et al. 1993), and can therefore be excluded from further consideration.

Using a simple two-component mixing model as in Rosman et al. (1993) allowed us to estimate that Eurasian pollution sources contributed 18% and 56% of the isotopic lead burden in saxifrage located, respectively, between 79 - 80° N and 80 - 82° N on northern Ellesmere Island (Figure 6.5). Further, because radiogenic lead is preferentially lost from rocks during weathering, soils often have $^{206}Pb/^{207}Pb$ values exceeding 1.20 with anthropogenic inputs arising from unweathered ore bodies being, as suggested in this study, less radiogenic; i.e. $\leq$ 1.19 (Shirahata et al. 1980; Gobeil et al. 1995). In particular, limited sediment core data from two ponds on Ellesmere Island (Douglas et al. 1994) indicate that preindustrial isotope ratios for this region of the High Arctic vary between 1.339 ± 0.018 and 1.318 ± 0.015. Therefore, it is important to note that our mixing model calculations or percent anthropogenic Eurasian lead in saxifrage must represent conservative estimates due to the almost certain entrapment of locally-derived lead in soil particles.

Although there is little evidence form either this study or in Chapter 5 for the transport to and deposition of excess Eurasian lead to southern (76 - 78° N) Ellesmere Island, some of this material, nevertheless, appears to have reached the central portion of the island (78 - 80° N). With the exception of sampling site No. 3, in all locations, including those between 78 - 80° N) (sites Nos. 2 and 4), the duplicate measurement having the higher lead concentration also exhibited the lower, i.e. more Eurasian, $^{206}Pb/^{207}Pb$ ratio (Figure 6.5). In those locations from central Ellesmere Island where the lead burdens are low, excess Eurasian lead therefore had the effect of producing a major shift in the subsequent isotopic signature of the samples.

Although there has been an overall decrease in the total supply of lead to the High Arctic (Boutron et al. 1991) as a result of the North American decline in use of leaded gasoline (Blais 1996), our biological work supports previous atmospheric (Sturges and Barrie 1989) and precipitation (Rosman et al. 1993) sampling in suggesting that Eurasian sources may now predominate as the major input pathway to this region. Ellesmere Island, therefore, despite being one of the most remote and unpopulated regions on the world, is free of neither the over (Chapters 14, 15) nor subtle (Chapters 1 – 5) by-products of human society.

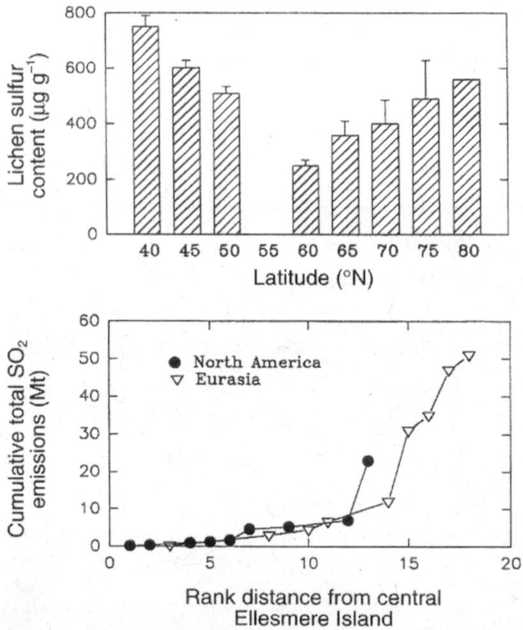

Figure 6.1. Upper panel: Latitudinal distribution of sulfur deposition in eastern Canada as measured by lichen accumulation. Data sources are listed in Chapter 5 (Figure 5.1). Lower panel: Cumulative total SO_2 emissions in relation to ranked distances from central Ellesmere Island (#1-18) of circumpolar source regions. Data are from Barrie et al. (1992).

Figure 6.2. Arctic haze observed at northern Ellesmere Island at beginning of the 1990 ski expedition.

Figure 6.3. Flowering purple saxifrage from southern Ellesmere Island near finish of 1990 ski expedition.

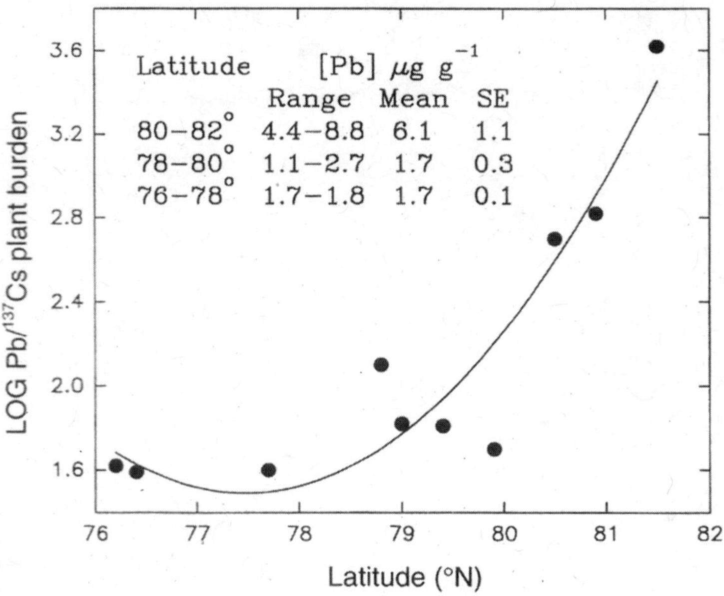

*Figure 6.4. Lead accumulation ($\mu g\, g^{1}$ dry wt) in plants (*Saxifraga oppositofolia*) collected from 10 locations along a 6° north-to-south transect on Ellesmere Island. Latitude explained 88% of the variation in $Pb/^{137}Cs$ plant burdens (solid regression line). Cesium-137 was measured on an intrinsic Ge counter which produced an overall error in counting rates of 12% as described in Chapter 1.*

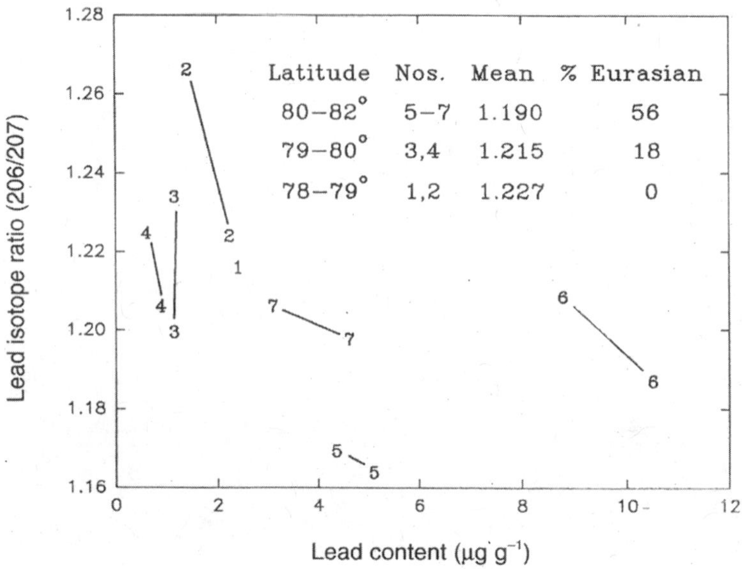

Figure 6.5. Lead isotopic composition (ratio of $^{206}Pb/^{207}Pb$) in saxifrage from central (sampling sites Nos. 1 and 2), north-central (sampling sites Nos. 3 and 4), and northern (sampling sites Nos. 5-7) Ellesmere Island. Isotopic signatures for American and Eurasian emissions average 1.23 and 1.16, respectively (Sturges and Barrie 1989), and were used to estimate the relative contribution of Eurasian sources to the plant Pb burdens. These are conservative estimates in that they assume no incorporation of locally derived soil particles characterized by an average Pb ratio of 1.33 (Douglas et al. 1994). Lead content explained 43% of the variability in isotope ratios.

PART II
WILDLIFE

CHAPTER 7

The Lake Hazen Trough: A Late Winter Oasis in a Polar Desert*

***Adapted from *Biol. Conserv.* 63:149-151; Abstract in Appendix II**

INTRODUCTION

Life in the High Arctic is distributed extremely nonrandomly. Whereas most of the terrestrial landscape is xeric, barren, and largely unproductive, certain other locations are alternatively comparatively lush and fertile. These latter regions are infrequent in occurrence, small in size, and generally represent no more than 1-2% of the total unglaciated terrain. Because most of these vegetated regions tend to be localized and geographically discrete from the surrounding biologically impoverished landscape, they can be referred to as High Arctic "oases" (Svoboda and Freedman 1981; Thomas and MacDonald 1986). Such oases are thought to exist through receiving more available solar energy and retaining more soil moisture than the surrounding polar desert, and most frequently occur in relatively flat coastal lowlands. One notable exception is the thermal oasis on northern Ellesmere Island (81° 45' – 82° 10' N, 68° 30' – 72° 45' W), which surrounds Lake Hazen, the largest lake within the Arctic Circle (Soper and Powell 1985).

Because Lake Hazen lies in a trough bounded abruptly by mountains (> 2000m) to the north and by a plateau (400 – 900m) to the south, it is effectively protected from incursions of unmodified coastal air (Corbet 1969). The result is a microclimate with higher temperatures and a prolonged frost-free season compared to the surrounding, more typical high-latitude, regions (Greely 1886; Soper and Powell 1985). This, together with the increased precipitation and consequent soil moisture, produces a lush vegetation which attracts abundant migrant wildlife (Greely 1886; Saville 1964; Saville and Oliver 1964; Nettleship and Mahr 1973; Gould 1988). Because of recognition of the uniqueness of the area, Lake Hazen was incorporated into one of Canada's most recent, and North America's most northern, protected reserve (England 1983; Environment Canada 1989).

The purpose of the present study was (a) to determine if the favorable summertime conditions in the Lake Hazen Trough are sufficient to maintain elevated abundances of non-migratory wildlife during periods of non-vegetative growth; and (b) to quantify the range of influence of the Hazen Oasis upon this biogeographic pattern during the late-winter period, at which time the region is subjected to possibly the coldest weather in all of North America (Soper and Powell 1985) with temperatures often below -30°C.

Survey Description

Although researchers heuristically believe polar oases like the one surrounding Lake Hazen to be characterized by higher abundances of wildlife, comparative investigations between such "oases" and surrounding unproductive "deserts" (*sensu* Soper and Powell 1985) are lacking. The present study, which undertook a traverse through the Lake Hazen Trough (Figure 7.1), was designed to contrast the notable paucity of life outside with the prominent abundance inside the Hazen Oasis. Distances traveled included 150, 124, and 145 km north, within, and south of the Lake Hazen Trough, respectively. Only Fielden (1877a,b), Greely (1886), MacDonald (1953) and Tener (1965) have previously published reports of wildlife observations in areas surrounding the Lake Hazen Trough.

Wildlife surveys in the remote High Arctic are customarily undertaken from rapidly moving aircraft despite knowledge of the biases, inconsistencies and inaccuracies implicit in such census techniques (e.g. Caughley 1974; Gasaway et al. 1985; Conroy et al. 1988). This situation is especially true with regard to smaller species of wildlife in their cryptic winter colouration. In contrast, the slow rate of progress of 11 (± 4 SD) km skied per 8 – 12 hr day (Appendix I) allowed opportunity for observation, sometimes with binoculars and spotting scope, of wildlife encountered along the way which might otherwise have been disturbed by the presence of motorized vehicles.

Detectability of animals with respect to perpendicular distance from the survey line is species-particular. Large and distinct animals such as muskoxen (*Ovibos moshatus*) or Peary caribou (*Rangifer tarandus*) can be easily spotted at distances occasionally exceeding 5 km on either side of the line of travel. Other animals have to be within only a hundred to a dozen meters to be spotted, however. Consequently, abundances can at best be expressed as only relative densities which underestimate the true values. During this part of the expedition, only three days were inclement enough to cause potential bias of the counts, which otherwise remained at a consistent level of effort. Because my purpose was to highlight the exceptional role of the Hazen Oasis in influencing wildlife abundance, and not to generate accurate and precise density estimates, the present methodology is considered to be accurate for this task (see also Chapter 8).

The influence of High Arctic oases upon non-migratory faunal abundance during the harsh winter and early spring periods is in general unknown, such investigations never in particular having been previously undertaken in the Lake Hazen trough. The present study, however, was initiated on March 30, 1990 at Cape Aldrich (the northernmost point of land in North America), reached Lake Hazen on April 20, 1990, and was terminated on May 6, 1990 when we began crossing high-altitude glaciers to the south of the oasis (Appendix I). This interval of time, and the fact that only non-migratory animals were observed, negates any confounding seasonal influences upon species abundance or distribution.

RESULTS AND DISCUSSION

Four species of mammals were observed during the survey: muskoxen, arctic hare (*Lepus arcticus*), arctic fox (*Alopex lagopus*), and Peary caribou. Arctic wolf (*Canis lupis*) were seen at the mouth of the Ruggles River on the south shore of Lake Hazen one week after we left the area (P. Wilkinson, pers. comm.). Rock ptarmigan (*Lagopus mutus rupestris*) and snowy owl (*Nyctea scandiaca*) were the only resident birds seen.

The abundance of mammalian and avian wildlife during the late winter was considerably greater within the vegetatively productive Lake Hazen Trough compared to the relatively depauperate desert landscape north or south of the oasis (Figure 7.2). The frequency of observed wildlife was positively correlated ($R_{Spearman} = 0.94$, $n = 4$ for data grouped according to daily frequency counts of 0, 1-3, 4-7, and >8 animals) with mean proximity to Lake Hazen itself. The vegetative base developed each summer surrounding Lake Hazen is therefore sufficient to provide continual support for animals during the long and particularly harsh winter. This is different from the situation thought to exist for the smaller Alexandra Fiord Oasis located further south on Ellesmere Island (Svoboda and Freedman 1981).

Benign systems such as the Hazen Oasis can be conceived of as being "open" (von Bertalanffy 1950) with a continuous exchange of matter and/or energy with the outlying, not so benign, environment. Such ecosystemic "centres of organization" (Steedman and Regier 1987; Chapter 8), though being relatively discrete in area, may therefore have a role far more important than would be suggested by their physical size alone (Svoboda and Freedman 1981).

Because High Arctic oases function similarly to other centres of organization such as temperate wetlands (France 2003) or lakeshore accumulations of woody debris (France 2006) in exporting "information" (Atlan 1985; France 2006), their discovery, identification and preservation are of utmost importance. The present survey suggests that levels of resident faunal abundance during the late winter/early spring do not approach background levels representative of polar desert environments (i.e. <3 animals observed per 11-km day of surface travel) until distances exceeding 50 km from the locus of Lake Hazen. Anthropogenic activity should obviously be restricted within this "nodal region" (see Noss and Harris 1986). Similar data documenting comparable zones of influence for other centres of organization (e.g. polynyas – Chapter 8) are critically needed in the present age of increasing development pressures within the High Arctic. Because oases are so biologically important, their candidacy as World Heritage Sites (von Droste 1985) or as International Biological Program Ecological Sites (Nettleship and Smith 1975) may be one approach to ensure their continual preservation in pristine conditions.

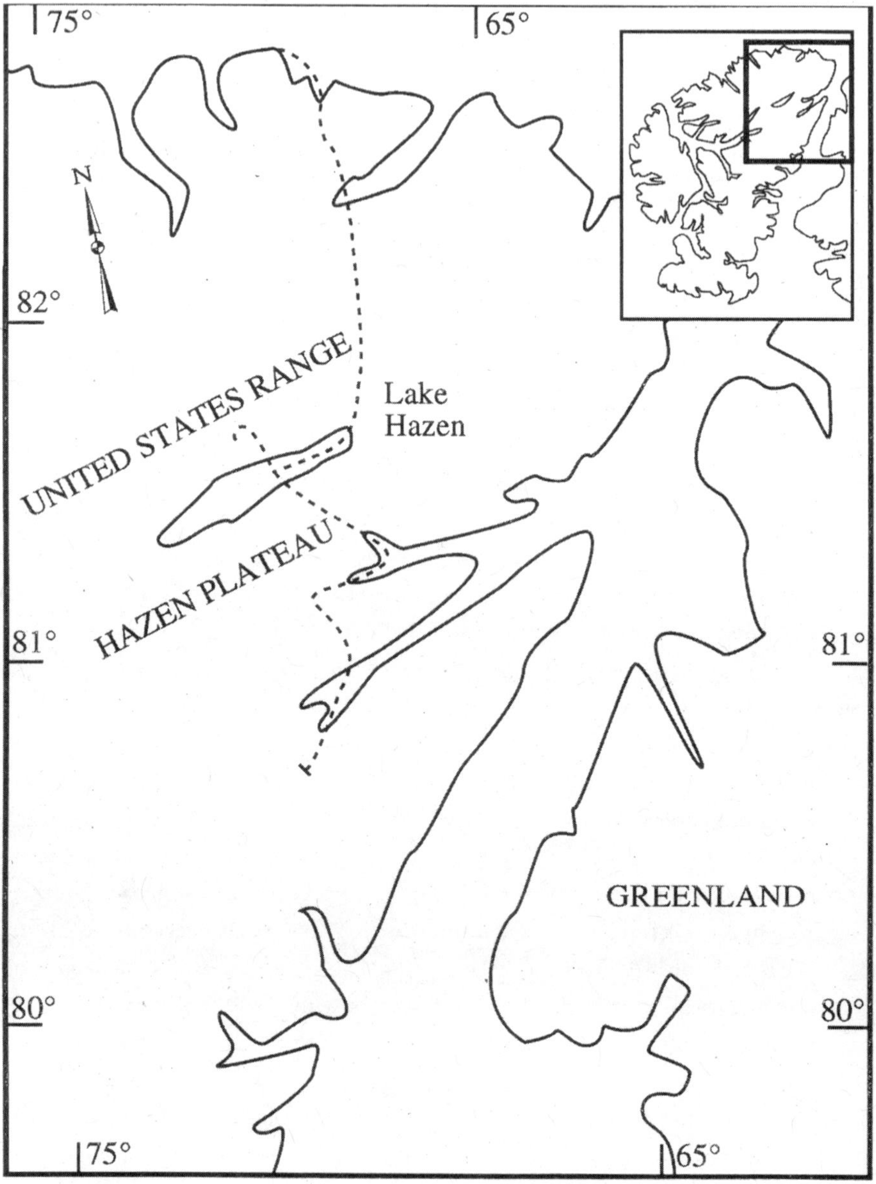

Figure 7.1. Location of the Lake Hazen Oasis and the 415 km expedition route on northern Ellesmere Island.

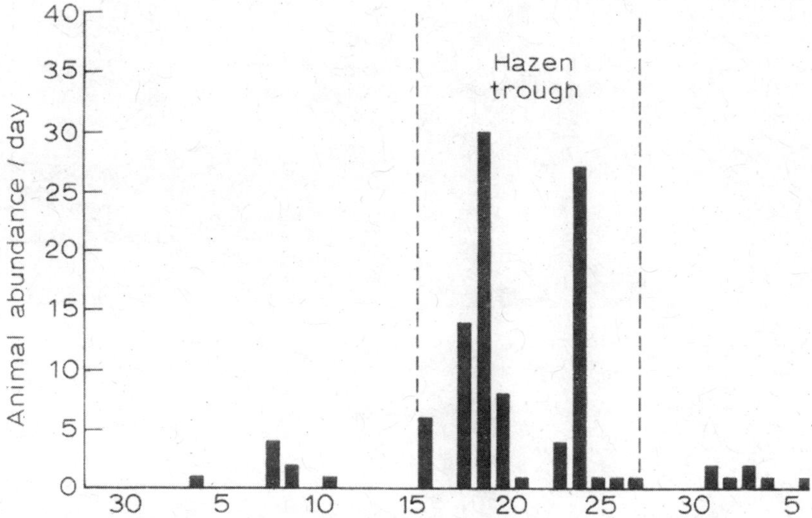

Figure 7.2. Chronology of wildlife (mammals and non-migratory birds combined as in Chapter 8) observed during the late winter/early spring traverse of 150, 124, and 145 km north, within, and south of the Lake Hazen Trough on northern Ellesmere Island. Average daily rate of progress (see Appendix I) was 11 km. Dates of April 21, 22 and 25 were spent on the ice of Lake Hazen itself.

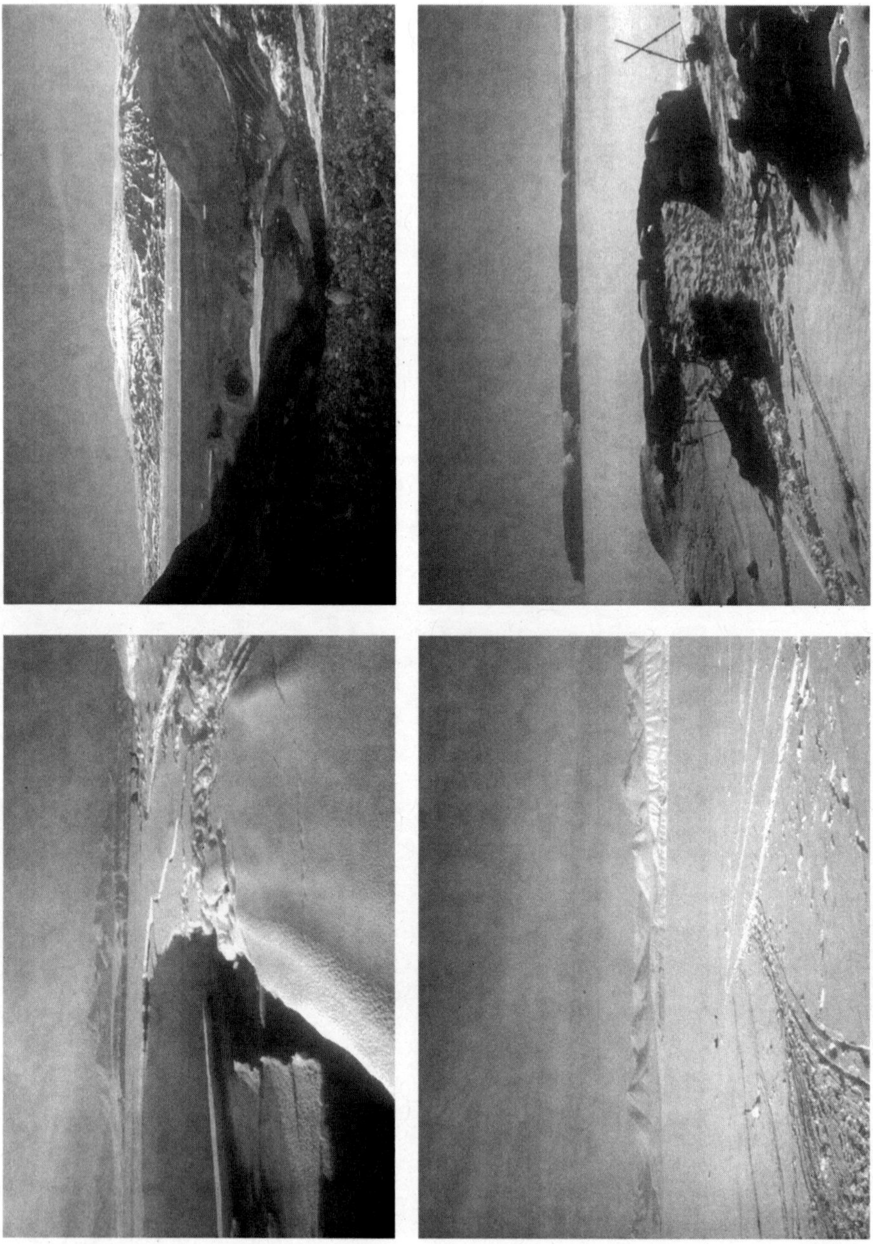

Plate 8. Wildlife oases on Ellesmere Island: Lake Hazen (bottom-left, top-left), Princess Marie Bay (bottom-right) and Makinson Inlet Head (top-right).

CHAPTER 8

Polynyas as Centres of Organization for Structuring the Integrity of Arctic Marine Communities*

***Adapted from *Conserv. Biol.* 6:442-446; Coauthor is M. Sharp**

INTRODUCTION

Polynyas are areas of open water surrounded by ice that are kept from freezing by the combined effects of wind, tidal fluctuations and upwellings (Nutt 1969; Sadler 1974; Dunbar 1981). Although the locations of many annually recurring polynyas in the High Arctic are well known (Smith and Rigby 1981), almost nothing is understood about aquatic processes that occur there (Stirling 1980; Stirling et al. 1981). Since the early explorers and whalers first penetrated the waters of northern Canada, the close association of biological activity with polynyas has been noted (Dunbar and Dunbar 1972). Today, recurring polynyas are recognized to be of critical importance in providing winter refuge for marine mammals (Kiliaan and Stirling 1978; Finley and Renaud 1980; Stirling 1980; Stirling et al. 1981) and as staging grounds for feeding and reproduction of migrant birds during the spring (Renaud and Bradstreet 1980; Prach et al. 1981; Brown and Nettleship 1981: McLaren 1982). Because the natural diet of Inuit and their PalaeoEskimo predecessors was completely dependent on animals, it is not surprising that the concentration of game around polynyas likewise attracted these top-level "predators" in both historic and prehistoric times (McMillan 1918; Schlederman 1980; Sutherland 1991). Unfortunately, knowledge of the biological significance of polynyas is still limited to only a number of anecdotal, qualitative accounts. The present study, which undertook a 550-km ground survey along the eastern coast of Ellesmere Island (latitude 80° 33' to 76° 7' N), documents the role of polynyas in quantitatively structuring the ecological integrity of marine artic communities.

Unlike previous studies, the present investigation compares by way of transect, the abundance and species richness of wildlife within polynyas to that of other areas during the spring when the polynyas are still bounded by vast expanses of stable sea ice. Many of the biases, inconsistencies, and inaccuracies implicit in aerial census techniques (Caughley 1974) were circumvented through use of a ground-based survey (Chapter 7), the first modern such investigation of extensive regions along the eastern coast of Ellesmere Island.

METHODS

The complete route undertaken by the Arctic Light Expedition is shown in Figure 8.1. We skied or walked a traverse of 550 km along the eastern coastline, broken into three sections when the expedition route moved inland atop icefields. Locations, time periods, and travel rates for each section were as follows: (A) Radmore Harbour (80° 33' N) to (B) Jokel Fiord (78° 47' N) from May 12 to June 6 at 16.0 ± 6.1 (SD) km per day (Appendix); (C) Makinson Inlet (77° 44' N) to (D) Bentham Fiord (77° 5' N) from June 15 to June 22 at 15.8 ± 7.1 km per day; and Jones Sound from (E) the Jakeman Glacier (76° 7' N) to (F) Grise Fiord (76° 8' N) from June 27 to July 4 at 15.7 ± 9.2 km per day. Three polynyas were encountered, one in each section: Flager Bay (79° 17' N, 75° 55' W; Schlederman 1980), Makinson Inlet (77° 31' N, 81° 40'W; Sadler 1974), and the North Water near King Edward Point (76° 8' N, 81° 5' W; McLaren 1982).

Our slow skiing rate of 1.6 km per hr allowed opportunity for observation, sometimes with binoculars and spotting scope, of wildlife that might have been disturbed by the presence of motorized vehicles. Sledging took place usually between 0200 and 1400 hr (the sun is up for the entire day at these latitudes). This entailed an entire day's careful observation at several sites along the ice edge. Species identification of birds was aided by Robbins et al. (1989) and consequently necessitated revision as better encounters occurred (for plovers for example). Land-birds (such as ptarmigans or buntings) and mammals (such as hares and foxes) are not considered in the present analysis.

While ground-based surveys typically provide more accurate estimates of population size than do aerial surveys which take place rapidly far above the surface, they are still by no means completely free of bias. Detectability of animals with respect to perpendicular distance from the survey line is species particular Large and distinct mammals such as seals or polar bears can be easily spotted at distances of about two to three km on either side of the line of travel. Vocal cues aid in sighting ranging gulls, although the width of visual identification is considerably less than that for large mammals, perhaps not more than one half to one km in total. Consequently, abundances can at best expressed only as relative densities, which underestimate the true values. During this part of the part of the expedition only four days were inclement enough to potentially bias counts, which otherwise remained at a consistent level of effort. Because our purpose was to highlight the exceptional role of polynyas in influencing the abundance and species richness of marine communities relative to background reference areas, and *not* to generate accurate and precise density estimates, we considered our methodology to be adequate for this task.

We were also interested in obtaining a rough estimate of the sphere of influence of the polynyas on eastern Ellesmere Island. Assuming an abundance level of 100% at the polynya itself, we determined the percentage of this value by counting the number of birds and mammals observed on each day either approaching or leaving the polynya. The spatial measures used were those of the mid-distance of each day's location from the polynya ice edge. These data were then grouped into six 50-km distances and averaged. Because of the progressive increase in wildlife abundance during the spring (due to increased seal haul-outs in warmer weather and northward avian migrations), data are expressed as proportional attenuations of abundance calculated polynya-specifically and then averaged for an overall trend.

Results and Discussion

Marine mammal abundance, avian abundance, and to a lesser extent, avian species richness, were all greater within the polynyas of eastern Ellesmere Island than in the generally depauperate and unproductive interlying stretches of coastline (Figure 8.2). Birds were most frequently concentrated along the ice edge in the open water. Seals were found to reach their greatest densities hauled out on the surrounding pack ice. No location preference was evident in the distribution of birds around the portion of the perimeters surveyed for either of the two discrete polynyas, Flager Bay and Makinson Inlet, which at the time were about 70 to 100 ha in area. Distribution of seals was more irregular, however, being influenced possibly by local topographic features that regulate the conditions of the fast ice near the polynyas. Species presence and abundance at King Edward Point on the southern tip of Ellesmere Island are of course influenced by the larger area of the North Water polynya in Baffin Bay (see Kiliaan and Sterling 1978; McLaren 1982).

Average abundances of birds and mammals at levels of 50% and 10% of the polynyal maxima were detected within 6 and 16 km, respectively, of the open water (Figure 8.3). Most significantly, faunal abundance did not reach background levels representative of the polar desert coastline until a distance of more than 30 km from the polynyas.

In the broadest sense, ecological "integrity" refers to the wholeness, diversity, and degree of connectiveness within a biological community, and can be viewed as an emerging property of ecosystems (France 1990, 2006; Karr 1990). Systems of high integrity have complex food webs (Ulanowicz 1990) and are often quite stable (Kay 1990). Benign systems that display high integrity such as polynyas are self-organized (Steedman and Regier 1990) and can be thought of as being "open" (von Bertalanffy 1950), with a continuous exchange of matter and/or energy with the outlying, not so benign, environment. Relatively small and discrete areas that provide essential conditions for maintaining ecological integrity may therefore have a role far more important than suggested by their size alone. Such locations can be referred to as ecosystemic "centres of organization" (Steedman and Regier 1987) or "nodes of diversity" (Noss and Harris 1986).

Because the survival of many species depends upon polynyas, when chance climatic conditions prevent the formation of open water, widespread mortalities occur (Barry 1968; Nettleship et al. 1980). Increasing offshore petroleum exploration within the High Arctic and the obvious attraction of persistently open stretches of water for boat transportation (Vanderwaag and Lamson 1989) have created concern about the environmental dangers these activities pose to polynyas (Sterling 1980; Sterling et al. 1981). If an accident such as that to the *Exxon Valdez* occurred within a polynya the potential for disaster would be profound. Not only would the local environment itself suffer but because the vast majority of wildlife for the entire region may be concentrated in a few polynyas during the winter and spring, the impact of an oil spill would be very far reaching.

Lack of quantitative data about the importance of polynyas in structuring arctic marine communities has previously prevented formulation of precise environmental recommendations. The present work indicates that polynyas exert a strong influence on the abundance and species richness of wildlife within a radius of 6 km, with effects being detected at distances of about 30 km. Anthropogenic activity should obviously be restricted with such "nodal regions" (Noss and Harris 1986). Similar data documenting comparative zones of influence for other arctic centres of organization such as lowland terrestrial oases (e.g. Chapter 7) are solely needed in an age of increasing development pressures within the High Arctic. Because polynyas are so biologically and historically important, their candidacy as World Heritage Sites (von Droste 1985) may be one means of ensuring their continual preservation in a pristine condition.

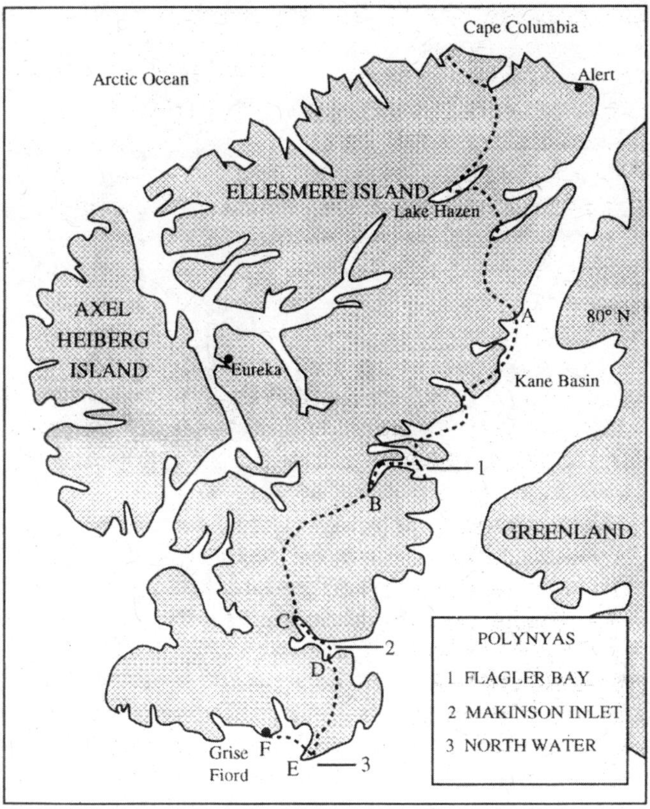

Figure 8.1. Location of polynyas (1,2,3) and expedition route (hatched line) along the coast (A to B, C to D, and E to F) of Ellesmere Island.

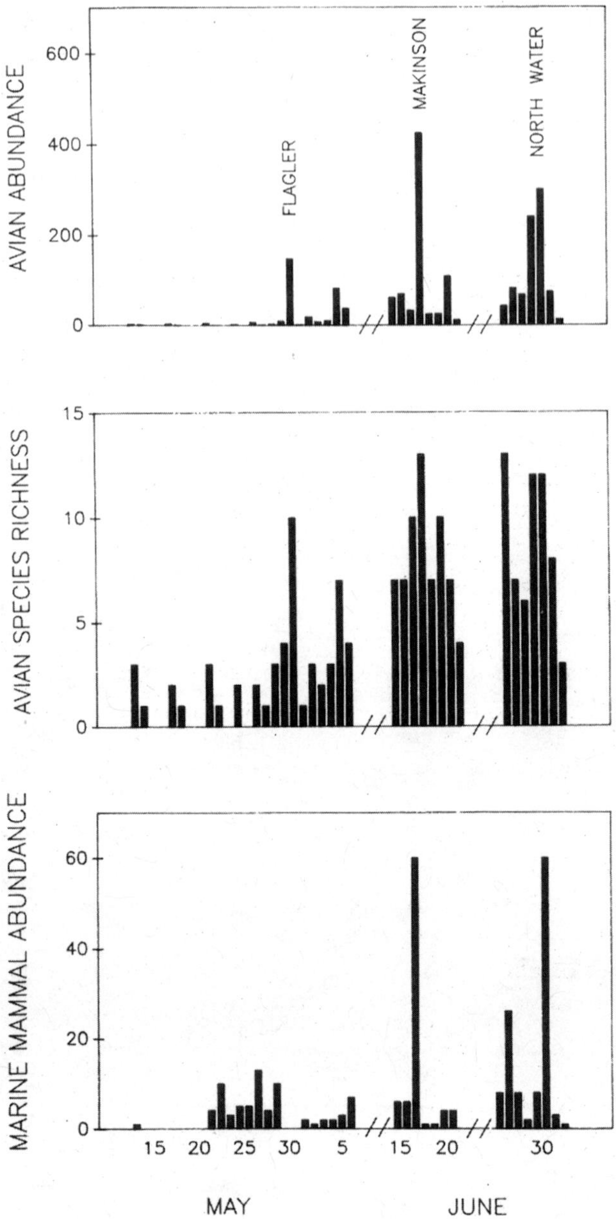

Figure 8.2. Chronology of wildlife observed during a 550-km ground traverse of the coast of Ellesmere Island during spring 1990. The three polynyas of Flager Bay, Makinson Inlet, and the North Water near King Edward Point were visited on May 29-30, June 17-18, and June 29-30, respectively. The symbols // denote breaks in the coastal traverse when the expedition route moved inland atop icefields.

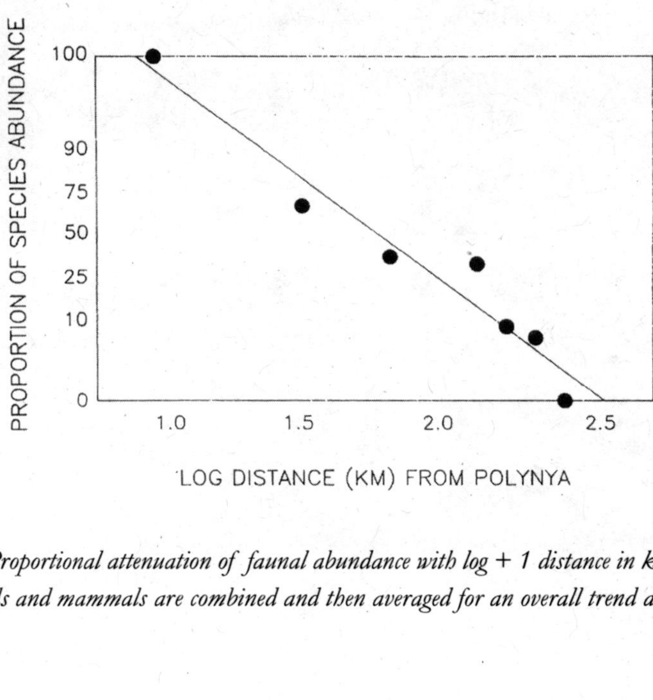

Figure 8.3. Proportional attenuation of faunal abundance with log + 1 distance in km from polynyas. Data for birds and mammals are combined and then averaged for an overall trend as described in the text.

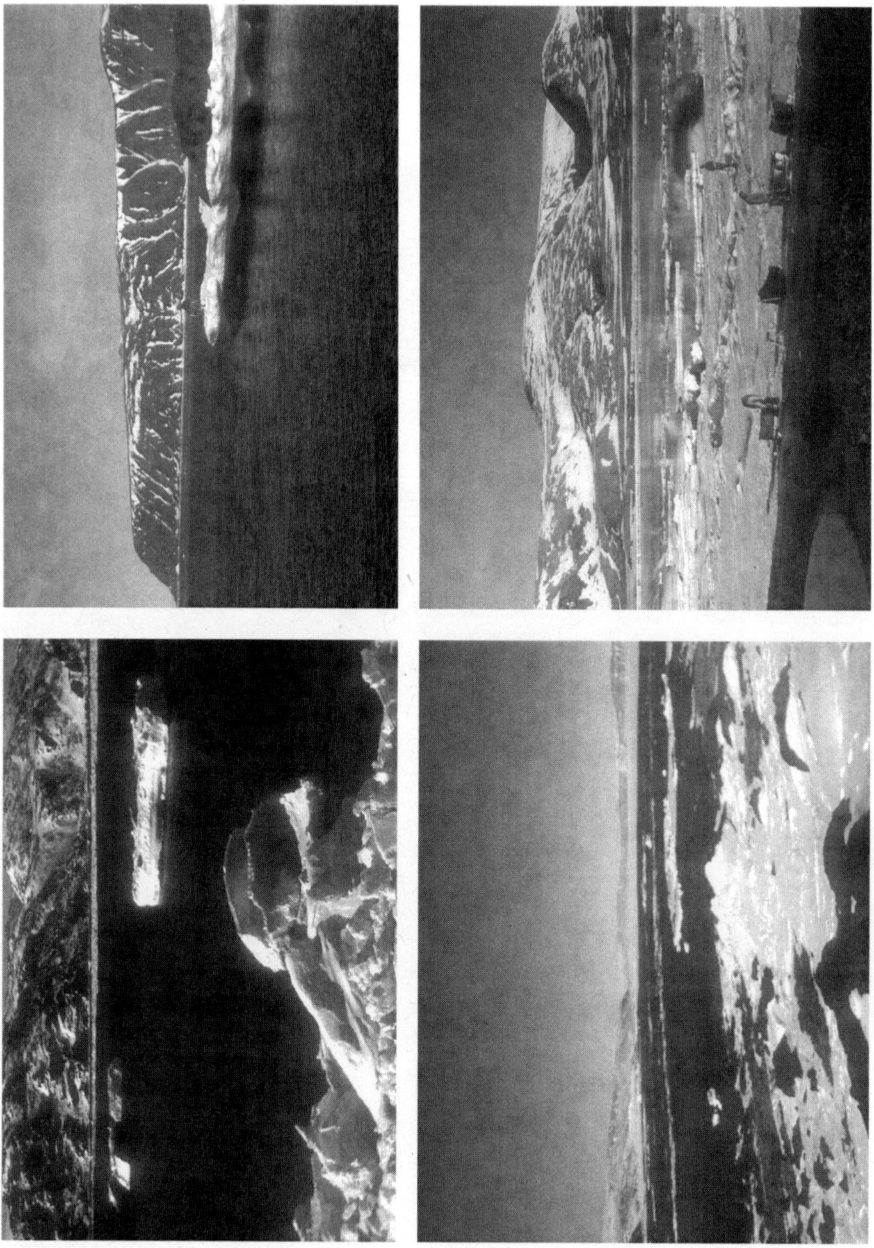

Plate 9. Ellesmere Island polynyas: Flager Bay (bottom-left, top-left), Makinson (bottom-right), and North Water (top-right).

CHAPTER 9

Comparison of Avian Species Richness and Abundance for Localized Oases on Ellesmere Island*

***Adapted from *Polarforschung* 61:147-152; Coauthor is M. Sharp**

Life in the High Arctic is nonrandomly distributed and often concentrated into relatively small and discrete localized areas which may have roles far more important than would be suggested by their physical size alone (e.g. Chapters 7 and 8). Until now, five such centres of diversity and biological productivity have been identified on Ellesmere Island (Figure 9.1). Three of these are the terrestrial oases of Lake Hazen (Soper and Powell 1985), Princess Marie Bay (Williams 1980), and Alexandra Fiord (Svoboda and Freedman 1981); and two of these are the polynyas of Flagler Bay (Schlederman 1980) and the North Water as it abuts the southeastern coast of Ellesmere Island (McLaren and Renaud 1982).

Oases exist as a result of geographically ameliorated microclimates or zones of high moisture which promote extensive floral productivity. Polynyas are areas of open water surrounded by ice which are kept from freezing by the combined actions of wind, tidal fluctuations, and upwellings. In both areas, avian communities are richer and more numerous here than in surrounding unproductive semi-desert or desert environments (e.g. Hussel and Holroyd 1974; Freedman and Svoboda 1982; Sterling and Cleator 1981; Chapters 7 and 8).

Although several aerial surveys of avifauna have been conducted on or along Ellesmere Island (e.g. McLaren and Renaud 1982; McLaren and McLaren 1982), due to biases, inconsistencies, and inaccuracies implicit in such studies (e.g. Diem and Lu 1960; Stott and Olson 1972; Caughley 1974; Conroy et al. 1988), ground-based surveys such as the present one are recognized to be the most accurate means of identifying and enumerating species. Limitations of accuracy and precision in aerial surveys can be specially severe for areas of high abundance in the High Arctic (McLaren 1982).

In the present study we document the avifaunal communities observed at four discrete locations on Ellesmere Island during the ski-traverse. We then compare our findings with those obtained from previous ground-based studies in order to generate for the first time a ranking of relative abundance for the most ubiquitous species inhabiting Ellesmere Island oases.

Locations (Figure 9.1) and dates of inventory (Appendix I) for the oases surveyed during the 1990 expedition are: Flager Bay Polynya (79° 17' N, 75° 55' W; May 29-30);

Makinson Inlet (77° 40' N, 81° 50' W; June 15-16); and the North water near King Edward Point (76° 08' N, 81° 05' W; June 29-30). These are the first biological observations from the Makinson Polynya, and the first putative identification of a nearby coastal oasis at the head of the same inlet (Chapter 12). Locations of the previously visited oases (Figure 9.1) are: Lake Hazen (81° 50' N, 71° 25' W; Saville and Oliver 1964; Nettleship and Mahr 1973; Gould 1988; Chapter 7); Princess Marie Bay (79° 25' N, 75° 45' W, Williams 1980); Flagler Bay Polynya (see coordinates above; Schlederman 1980; Chapter 8); and Alexandra Fiord (78° 53' N, 75° 55' W; Freedman and Svoboda 1982; Chapter 13). All were inventoried during June and July.

Seven, 16, 12, and 9 species were respectively observed (from north to south) in the Flagler Bay Polynya, coastal flats at the head of Makinson Inlet, the Makinson Polynya, and the North Water near King Edward Point (Table 9.1). In total, 1266 birds of 23 species were identified and categorized as being "Very Abundant", "Abundant", "Common", "Rare", and "Present" based on numerical divisions of 150, 75, 30, 10 and 1.

By synthesizing these data with those obtained previously for other biologically productive regions on Ellesmere Island (Table 9.1), we established the rank ordering of overall relative abundance of the 10 most ubiquitous and dominant species for (a) polynyas: Black Guillemots > Thick-billed Murres > Common Eiders > Arctic Terns > Oldsqaws > King Eiders > Glaucous Gulls . Kittiwakes > Ruddy Turnstones . Red-throated Loons; and for (b) coastal and thermal oases: Common Eiders > Black Guillemots > Glaucous Gulls > Snow Buntings > Baird's Sandpipers > Rock Ptarmigan > Snow Geese > Oldsqaws > Red Knots > King Eiders.

This ranking information can be used to correct some previous misconceptions made about avian diversity and relative abundance obtained from the more commonly undertaken aerial surveys; for example, the belief of McLaren and McLaren (1982) that "in the High Arctic common eiders are considerably less abundant than king eiders."

The present results can also be placed in broader context in relation to the total richness and abundance of all birds observed during the 96-day, 1300-km crossing of Ellesmere Island (Table 9.2), possibly the longest ground survey of wildlife undertaken in the High Arctic. Here, the importance of the four visited oases can be clearly seen in that over half of all birds observed and almost all species that were identified occurred within these few very small regions (which in fact comprise less than 3 to 8% of the total expedition distance and observation time). Continued recognition and documentation of such "centres of organization" (*sensu* Steedman and Regier 1987) or "nodes of diversity" (sensu Noss and Harris 1986) are solely needed in an age of increasing development pressure within the High Arctic. Because polynyas and terrestrial oases are so biologically (Chapters 7 and 8) and historically (Schlederman 1980; P. Sutherland, National Museum of Canada, pers. comm.) important, their candidacy as World Heritage Sites or National parks should be seriously considered.

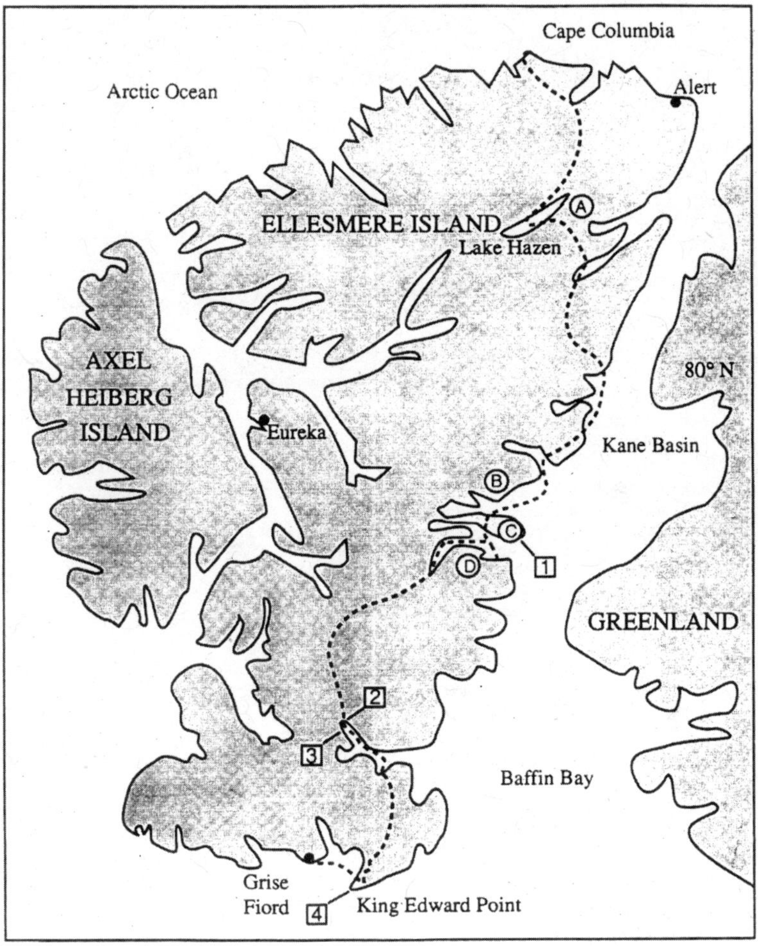

Figure 9.1. Expedition route (dashed line) and location of oases on Ellesmere Island. Letters indicate previously visited oases: A = Lake Hazen, B = Princess Marie Bay, C = Flagler Bay Polynya, and D = Alexandra Fiord. Numerals denote oases surveyed in 1990 during the Arctic Light Expedition: 1 = Flagler Bay Polynya, 2 = Makinson Inlet Head, 3 = Makinson Inlet Polynya, and 4 = the North Water.

Table 9.1. Avian inventory for seven oases on Ellesmere Island ranked; VA = very abundant, A = abundant, C = common, R = rare, P = present. (1) Savile & Oliver 1964, Nettleship & Mahr 1973, Gould 1988; (2) Williams 1980; (3) This study; (4) Schlederman 1980; (5) Freedman & Svoboda 1982: R = regular, B = breeding confirmed, other species present but no evidence of breeding; (6), (7), (8) This study. Species identifications in Table 9.2; (10) Species not found during this study.

	HAZEN OASIS (1)	PRINCESS MARIE BAY (2)	FLAGLER POLYNYA (3)
VA	—	Common Eider	—
A	—	Black Guillemot	Common Eider
C	—	Glaucous Gull Red Knot Baird's Sandpiper Rock Ptarmigan	—
R	—	Snow Goos Oldsquaw	Ruddy Turnstone
P	Snow Bunting (9) Northern Redpoll Long-tailed Jaeger Arctic Tern Rock Ptarmigan Oldsquaw Snow Goose	Red-throated Loon Brent Goose King Eider Gyrfalcon Ringed Plover (10) Golden Plover Sanderling (10) Purple Sandpiper (10) Ruddy Turnstone Long-tailed Jaeger Ivory Gull Arctic Tern Dovkie (10) Snowy Owl Northern Raven Lapland Longspur (10)	Snow Bunting Glaucous Gull Northern Raven Ivory Gull King Eider

	FLAGLER POLYNYA (4)	ALEXANDRA FIORD (5)	MAKINSON HEAD (6)
VA	Common Eider	—	—
	Oldsquaw		
	Black Guillemot		
A	King Eider	—	—
C	Glaucous Gull	Common Eider („R“)	Snow Bunting
	Arctic Tern	Glaucous Gull („R“)	
R	Dovekie	Snow Bunting (B)	Glaucous Gull
			Snow Goose
P	Long-tailed Jaeger	Northern Redpoll (B)	Oldsquaw
	Snow Bunting	Baird's Sandpiper (B)	Baird's Sandpiper
	Thick-billed Murre	Arctic Tern (B)	King Eider
		Lapland Longspur (B)	Brent Goose
		Rock Ptarmigan (B)	Ruddy Turnstone
		Oldsquaw (B)	Common Eider
		Snow Goose (B)	Rock Ptarmigan
		Parasitic Jaeger (B)	Ivory Gull
		Red-throated Loon	Snowy Owl
		King Eider	Long-tailed Jaeger
		Gyrfalcon	Parasitic Jaeger
		Ringed Plover(10)	Ruddy Turnstone
		Ruddy Turnstone	Rough-legged Hawk
		Purple Sandpiper (10)	
		Long-tailed Jaeger	
		Thayer's Gull	
		Ivory Gull	
		Snow Owl	
		Horned Lark (10)	
		Northern Raven	

	MAKINSON POLYNYA (7)	NORTH WATER (8)
VA	Arctic Tern	Thick-billed Murre
		Black Guillemot
A	—	—
C	Oldsquaw	Kittiwake
	King Eider	Glaucous Gull
	Common Eider	Common Eider
R	Snow Bunting	Brent Goose
	Red-throated Loon	Snow Bunting
P	Red Knot	Ivory Gull
	Glaucous Gull	Gyrfalcon
	Rock Ptarmigan	
	Northern Raven	
	Parasitic Jaeger	
	Ruddy Turnstone	

Table 9.2. Species identification and total number of birds observed for both oases and non-oases regions during the 96 day (March-July), 1300km Trans-Ellesmere Island Expedition.

Larus hyperboreus	Glaucous Gull	386
Pagophila eburnea	Ivory Gull	355
Somateria mollissima	Common Eider	228
Cepphus grylle ultimus	Black Guillemot	220
Sterna paradisaea	Arctic Tern	200
Plectrophenax n. nivalis	Snow Bunting	191
Uria l. lomvia	Thick-billed Murre	170
Clangula hyemalis	Oldsquaw	80
Somateria spectabilis	King Eider	73
Rissa t. tridacyta	Black-legged Kittiwake	60
Lagopus mutus rupestris	Rock Ptarmigan	42
Branta bernicla	Brant Goose	34
Arenaria i. interpres	Ruddy Turnstone	30
Chen hyperborea	Snow Goose	29
Calidris c. canutus	Red Knot	24
Corvus corax principalis	Common Raven	17
Erolia bairdii	Baird's Sandpiper	13
Gavia stellata	Red-throated Loon	12
Charadrius h. hiaticula	Common Ringed Plover	12
Stercorarius parasiticus	Parasitic Jaeger	9
Larus thayeri	Thayer's Gull	9
Falco rusticolus canicans	Gyrfalcon	4
Pluvialis squatarola	Black-billed Plover	4
Nyctea scandiaca	Snowy Owl	3
Stercorarius longicaudus	Long-tailed Jaeger	1
Phalaropus fulicaria	Red Phalarope	1
Buteo lagopus s. johannis	Rough-legged Hauwk	1

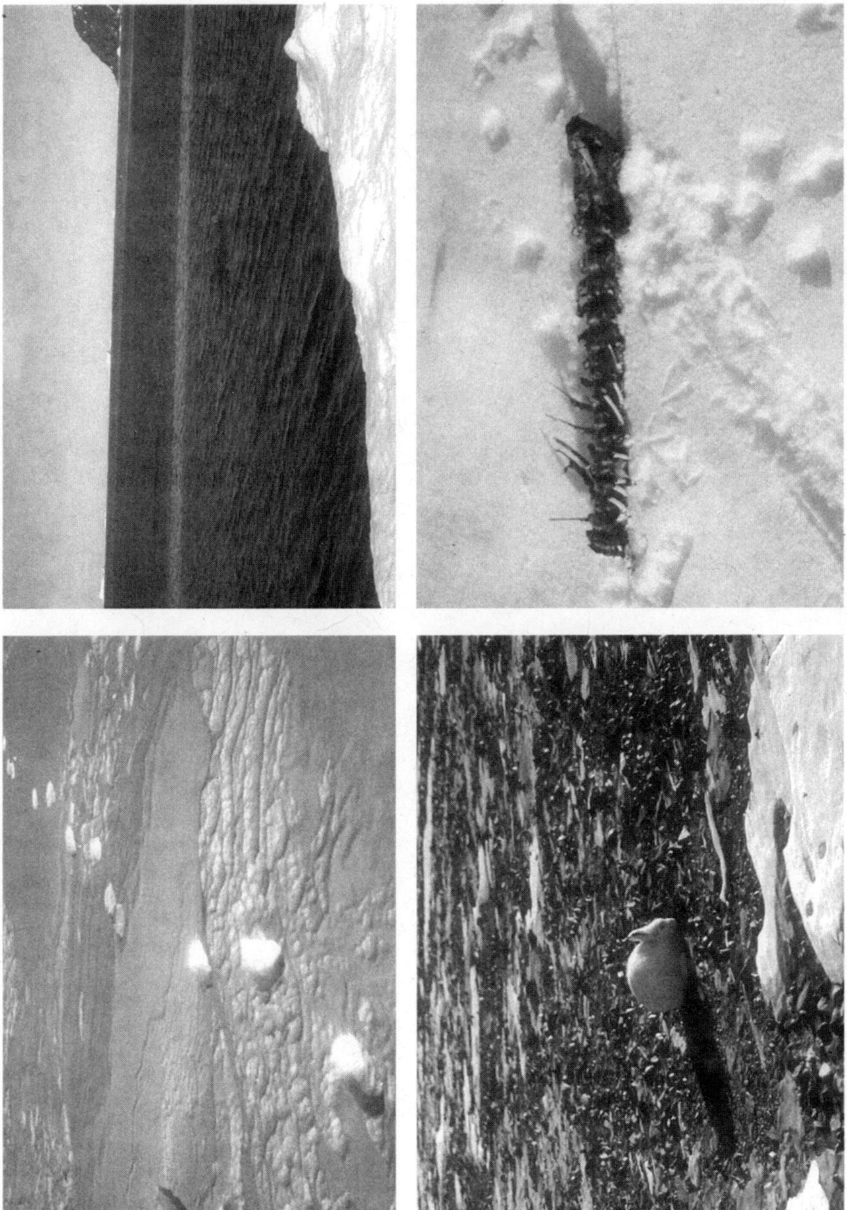

Plate 10. Wildlife and wildlife traces on Ellesmere Island: grazing Arctic Hare, Polar Bear tracks, seal backbone from Polar Bear attack, and swimming Walrus.

CHAPTER 10

Non-sterility of High Arctic Icefields on Ellesmere Island

Icefields comprise about 20% of the 76,600 sq miles surface area of Ellesmere Island, the ninth largest and, next to Greenland, most northerly island in the world. It has been customary to regard such ice-covered regions as being devoid of avian life: e.g. "Such areas [on eastern Ellesmere Island] are, of course, entirely destitute of breeding birds and may be disregarded in a further discussion of Arctic avifauna" (Soper 1940); and "...where a dead glacier descended from the [Spitzbergen] ice-cap" (Bateson and Plowright 1959). The first discovery of Ivory Gull (*Pagophilia eburnean*) breeding colonies high atop inland icefields in North America (Frisch and Morgan 1979) by geologists (biologists have rarely ventured into these incredibly remote regions) has challenged the concept of the sterility of the ice-cap.

During the 96-day, 1300-km Ellesmere Island ski expedition we discovered an additional 7 Ivory Gull colonies (Chapter 11). We also found the exposed rock nunataks emerging from the glacial ice to provide suitable nesting sites for both Rock Ptarmigan (*Laopus mutus rupestris*) (3 observed) and Snow Buntings (*Plectrophenax n. nivalis*) (14 observed). Additionally, both a Long-tailed Jaeger (*Stercorarius longicasudus*) and Northern Raven (*Corvus corax princepalis*) were seen in flight heading up glaciers. In all cases, the icebound seacoast or open water of polynyas, were most of the avifauna customarily resides (Chapters 8, 9, 12 and 13), were between 15 to 30 km distant from where these glacier birds were sighted.

The location of Ivory Gull colonies far from their feeding grounds on the coast is thought to be a 'strategy' to curtail competition from Glaucous Gulls (*Larus hyperboreus*) (Bateson and Plowright 1959), or more probably, to reduce egg/fledgling predation by Arctic Foxes (*Alopex lagopus*) (Larson 1960), the latter being capable of completely extirpating colonies found proximal to the coast (e.g. Dalgety 1932; MacDonald 1976). In this light, and given the rarity of Ivory Gulls in arctic North America (Thomas and MacDonald 1987), we were somewhat apprehensive by our sightings of several Arctic Foxes engaged in rapid and unidirectional transit atop the icefields. The possible luring of such effective predators from their typical lowland habitats to elevations of over 600m, and their potential impacts upon the once thought to be completely "secluded" and "sequestered" Ivory Gull colonies, is an unknown interaction meritorious of future study.

Plate 11. Locations of wildlife encounters: island at Alexandra Fiord, scene of aerial mobbing of Gyrfalcon (bottom-left); Makinson Inlet Head, scene of Rough-legged hawk sighting (top-left); and Prince of Wales Icecap, scene of bird and Artic Fox sightings (bottom-right, top-right).

CHAPTER 11

Newly Reported Colonies of Ivory Gulls on Southeastern Ellesmere Island*

***Adapted from *Arctic* 45:306-307; Abstract in Appendix II; Coauthor is M. Sharp**

Since the arctic explorations of the 19th century, when sightings of the near legendary Ivory gull (*Pagophila eburnean*) were notable events, questions concerning the location of their breeding sites have often been raised (Blomqvist and Elander 1981). Part of the difficulty in locating these sites arises not only from their remoteness and inaccessibility, but also from the fact that although widely distributed, the Ivory gull is regarded as extremely "rare" (World Wildlife Fund, Committee on the Status of Endangered Wildlife in Canada, June 1989), with an overall population of about 2400 adults throughout the eastern Canadian High Arctic (Thomas and MacDonald 1987). Until 1977, only a single breeding colony had been located in Canada (MacDonald 1976). Since then, the discovery of new Ivory gull colonies has been of continual interest, as have observations of general biology at locations distant from the colonies (Brown et al. 1974; Divoky 1976; Renaud and McLaren 1982; Orr and Parsons 1982.

Although Ivory gull colonies have been reported from lowland areas such as the Brodeur Peninsula of Baffin Island (two colonies, 48 birds–Reed and Dupuis 1983; eight additional colonies, about 570 birds –Thomas and MacDonald 1987), southern Ellesmere Island (one colony, about 300 birds –Thomas and MacDonald 1987), and northern Devon Island (one colony, 7 birds–Thomas and MacDonald 1987), it now appears that a major habitat of Ivory gull breeding colonies is the eroded nunataks that emerge from the many icefields of the High Arctic. These may be situated at elevations exceeding 1000m above sea level and dozens of kilometers inland. Because of the long-held belief among biologists in the sterility of such glaciated regions for bird life (e.g. Soper 1940; Chapter 10), it is not surprising that a geologist rather than an ornithologist first identified the importance of these regions for Ivory gulls in North America (five colonies, about 175 birds, on southeastern Ellesmere Island–Firsch and Morgan 1979; four colonies, about 91 birds, on northeastern Devon Island–Firsch 1983). Since then, Thomas and MacDonald (1987) have reported an additional ten colonies (approximately 12 to 300 birds present at each) around the Makinson Inlet region of southeastern Ellesmere Island. The present study reports the finding of seven further colonies, comprising a total of approximately 330 Ivory gulls, located during a ground traverse of 319 km of glaciers and upland icefields on southern Ellesmere Island during spring 1990.

From June 6 to 15, we crossed the western side of the Prince of Wales Icefield north of Makinson Inlet (Jokel Fiord to the Split Lake Glacier), and from June 22 to 27 we crossed the Manson Icefield (Figure 11.1) south of Makinson Inlet (Bentham Fiord to the Jakeman Glacier).

Most previous studies describing locations of Ivory gull colonies were conducted with aerial surveys and gave admittedly only "very approximate" estimates of gull abundance (e.g. Fisch and Morgan 1979), which may be far from absolute numbers. Because considerable time is needed to survey gull breeding sites quantitatively (e.g. Snell 1989), single aerial visual estimates are not adequate to measure the number of gulls in individual colonies. Likewise, because Ivory gulls are very sensitive to disturbance, the presence of aircraft of motorized vehicles can influence their detectability (V. Thomas, Zoology Dept., Univ. Guelph, pers. Comm., 1991). The average rate of progress of the expedition through the icefields of southern Ellesmere was 19 km per 8-12 hr day of travel on skis (Appendix I) which permitted detailed observation of any colonies located along the way.

No colonies of Ivory gulls were observed on the 28 nunataks we passed along the 170 km traverse of the western edge of the Prince of Wales Icefield, north of Makinson Inlet. We did, however, observe wandering birds: two near the Flagler Bay polynya, four in Jokel Fiord near the foot of the Stygge Glacier, and one just before descending to the head of the Makinson Inlet. The nine colonies previously located on the Prince of Wales Icefield (Frisch and Morgan 1979; Thomas and MacDonald 1987) were over 70 km distant from our transect, toward the eastern coast.

Colonies of Ivory gulls were discovered on 7 of the 53 nunataks we passed during the 149 km traverse of the Manson Icefield, south of Makinson Inlet (Figure 11.1). Counts of adult birds ranged from 20 to 90 individuals per nanatak (Table 11.1).

The present study supports Thomas and MacDonald's (1987) contention that southeastern Ellesmere Island may contain a significant component of the total Ivory gull breeding stock in the eastern Canadian High Arctic. The location of numerous colonies here may be a result of their close proximity to the polynyas of Makinson Inlet and the North Water (Brown and Nettleship 1981; Chapters 8 and 9). This may also explain why Ivory gulls locate their colonies on the east, but now west, side of the Prince of Wales Icefield, farther to the north.

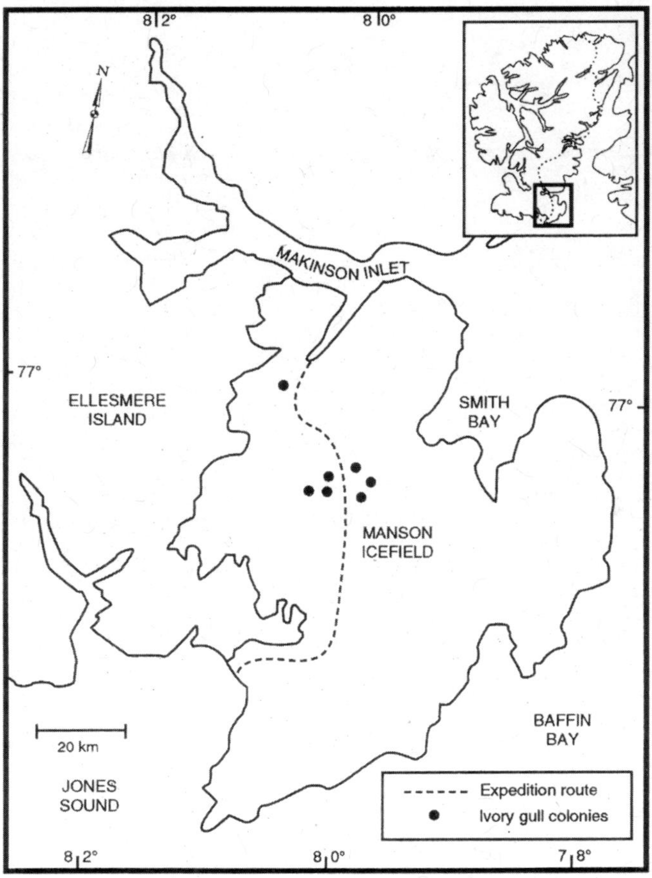

Figure 11.1. Location of expedition route and 7 Ivory gull colonies found on the Manson Icefield of southeastern Ellesmere Island.

Table 11.1. Location of Ivory gull colonies and number of birds observed on southeastern Ellesmere Island in spring 1990.

Map identification[1]	UTM[2]	No. of birds
Vendom Fiord 49D	NR 1048	24
Clarence Head 39B	NR 2725	28
Clarence Head 39B	NR 3233	20
Craig Harbour 49A	NR 2823	28
Craig Harbour 49A	NR 1525	70
Craig Harbour 49A	NR 1825	90
Craig Harbour 49A	NR 2315	70

[1]Energy, Mines and Resources Canada, District of Franklin, N.W.T. Scale 1:250 000.
[2]Universal Transverse Mercator Grid Reference.

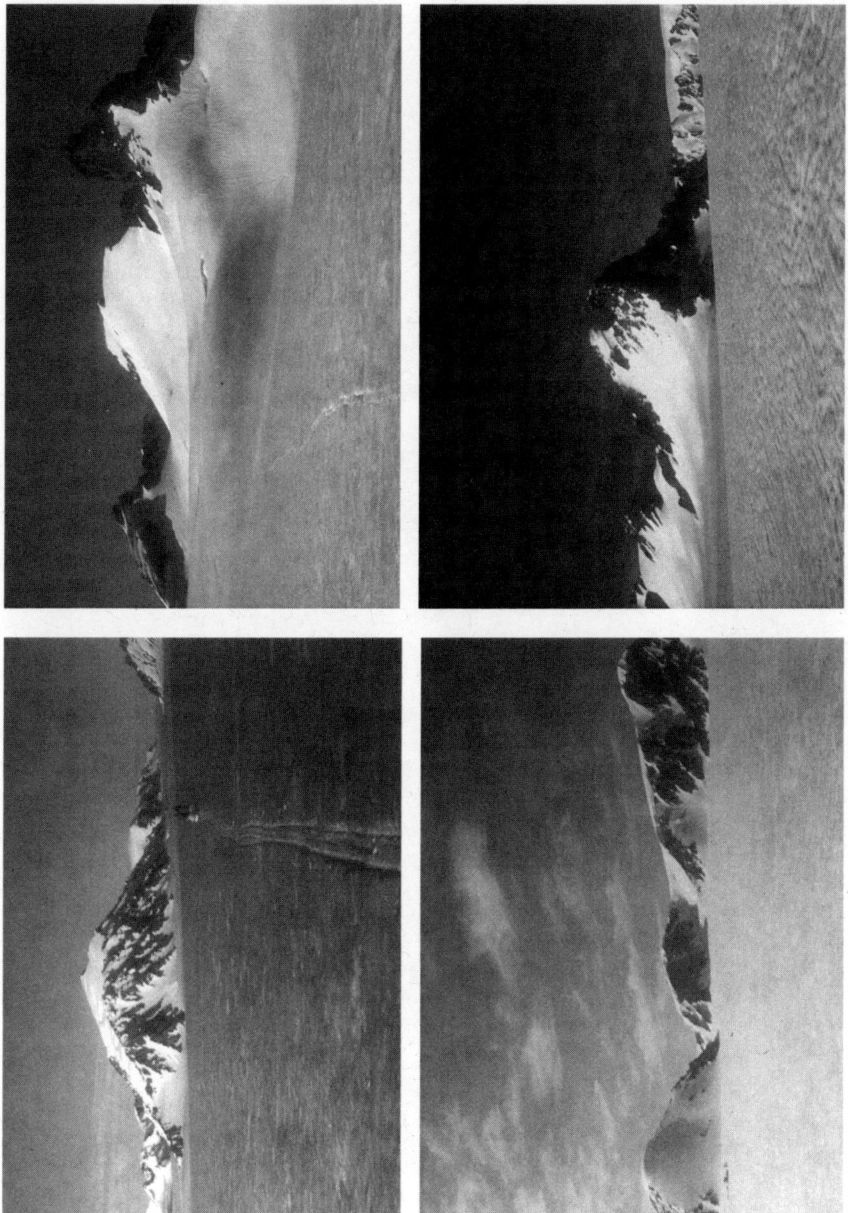

Plate 12. Location of discovered Ivory gull colonies on Manson Icefield nunataks on southern Ellesmere Island.

CHAPTER 12

First Record of the Rough-legged Hawk, *Buteo lagopus*, from Ellesmere Island, Northwest Territories*

***Adapted from *Can. Field Natur.* 106:511-512; Abstract in Appendix II; Coauthor is M. Sharp**

Only three diurnal raptors are thought to breed within the archipelago of the Canadian Arctic: the Gyrfalcon (*Falco rusticolus*), the Peregrine Falcon (*Falco peregrinus*), and the Rough-legged hawk (*Buteo lagopus*). Knowledge of the latter's distribution is incomplete (Bray 1943; Porsild 1951; Sutton and Parmelee 1956; Fraser 1957; Godfrey 1966, 1986). Rough-legged Hawks have not been recorded from Greenland (Sutton and Parmelee 1956; Burnham and Mattox 1984) or the Canadian islands of the High Arctic, specifically Devon, Cornwallis, Ellef Ringnes, Amund Ringnes, and Axel Heiberg (Godfrey 1986; Pattie 1990). Similarly, this species has not been documented for Ellesmere Island (Fielden 1877; Greeley 1886; MacMillan 1918; Soper 1940; MacDonald 1953; Godfrey 1953; Parmelee and MacDonald 1960; Williams 1980), despite detailed lists being produced there for favorable habitats for avifauna (Savile and Oliver 1964; Nettleship and Mahr 1973; Freedman and Svoboda 1982: Schlederman 1980; Gould 1988).

While undertaking a ski-traverse of Ellesmere Island our expedition identified a single adult Rough-legged Hawk (in pale phase) at the head of Makinson Inlet (Chapter 9) which bisects the icefields of the southeastern corner of the island (Fig. 11.1). At 77°40' N, this sighting extends the northern distribution of the Rough-legged Hawk in the eastern High Arctic by about 500 km. The hawk was observed circling the same localized area on two separate occasions during June 16 and 17, suggesting it was not engaged in a rapid fly-through but was at least a temporary resident.

The Makinson Inlet region is very remote, not being fully explored until the 1930s (Haig-Thomas 1939). The particular location in which the hawk was observed is an area of about 100 km^2, being bounded to the east by the Prince of Wales Icefield rising to a height of 1300 m, and to the north and south by the Split Lake and Hook Glaciers which descend to sea level (Vendom Fiord Map 1:250,000 49D). This region surrounds a large lake and is separated from the coast on the west by a range of hills exceeding 300 m. Located only 10 km away over the still frozen sea ice of late June, is the seldom observed Makinson Polynya (Sadler 1974). In comparison with the encircling polar desert, the enclosed putative "oasis" contains a particularly diverse and productive early spring avian community including Snow

Buntings, Baird's Sandpipers, Red Knots, Ruddy Turnstones, Rock Ptarmigan, Parasitic and Long-tailed Jaegers, and Common Ringed Plovers (Chapter 9). Although a pair of Snowy Owls was observed nesting at the site, the duration of occupation by the solitary Rough-legged Hawk is unknown.

CHAPTER 13

Aerial Mobbing of a Gyrfalcon by Glaucous Gulls*

***Adapted from *J. Raptor Res.* 26:269.**

On June 1-2, 1990 while camped at a small island in Alexandra Fiord on Ellesmere Island (78°54' N, 75°45' W) near a lowland oasis (Freedman and Svoboda 1982), I observed a male Gyrfalcon (*Falco rusticolus*) quartering back and forth in front of a cliff where the female was nesting. The orographic lift of the gyrfalcon was interrupted by a group of three Glaucous Gulls (*Larus hyperboreus*) on two separate occasions, each lasting several minutes. During the first encounter, the gulls, which initiated the engagement, acted in unison, each gull taking turns diving at the Gyrfalcon, while the others circled close by calling loudly. The Gyrfalcon, in an attempt to fight the gulls, swiveled about, talons extended, occasionally even doing so while completely inverted. After returning to the nest, the Gyrfalcon ignored the gulls until after 10 min it tried to make a kill of a single passing bird. The panicked avoidance by the Glaucous Gull consisted of dropping vertically to several meters above the pack ice and calling loudly. Within a few moments, two gulls arrived and the three dissuaded the Gyrfalcon from hunting. The falcon then returned to the nest site where it stayed for the next hour of observation, disregarding nearby gulls.

Gyrfalcons commonly take prey in the air (White and Weiden 1966; Bengston 1971) and customarily prey upon seabirds (e.g. Dementiev and Gortchakovskaya; Poole and Boag 1988) including Glaucous Gulls (White and Cade 1971). Similar mobbing of Gyrfalcons by Common Raven (Corvus corax) has been noted previously (Jenkins 1978). Wooden (1980) also observed perched Gyrflacons struck by passing Arctic Terns (*Sterna paradisaea*) which, however, never grouped to drive the raptor away as in the present encounter with the larger Glaucous Gulls. This seemingly paradoxical behavior of self-endangerment by mobbers may be necessary to "convince" the predator that their threat is real (Sordahl 1990). The result is that both ravens (Jenkins 1978) and Glaucous Gulls (this study) can be ignored even when they fly directly beneath or over an occupied Gyrfalcon eyrie.

PART III

Overt Pollution

CHAPTER 14

Garbage in Paradise*

***Adapted from *Nature* 355:504.**

The High Arctic is the ground "floor of creation" (Lopez 1987), the land having only recently emerged from its protective cradle of ice. In spring 1990, I was a member of a 1,300-km ski-traverse of Ellesmere Island (76 – 84° N), the northernmost land in North America. We came to realize that if curiosity in paradise was the original sin, then certainly the cavalier jettison of human garbage must be a close second.

Far from being an untainted wilderness, Ellesmere Island is riddled with the detritus of human passage and occupation. We encountered 208 objects, listed in Table 13.1, or one piece of garbage every 4 km of non-glaciated surface travelled. Yet Ellesmere Island is remarkably isolated, containing only about 100 permanent residents although it is the size of Great Britain.

No single problem of garbage is more pervasive in the Arctic than that of aircraft gasoline drums discarded by scientists studying the natural environment. A further irony is that the single largest collection of garbage we encountered was identified as belonging to a geologist who had written an article arguing against establishment of a national park on Ellesmere Island because it would attract too many tourists to an "environment [which] has exceptional limits in sustaining and suffering human use, even by the most modest southern standards." The Polar Continental Shelf Project, the umbrella government research organization in the Canadian Arctic, must enforce a "garbage in – garbage out" policy to all researchers relying upon its support.

The High Arctic is extremely unforgiving when it comes to absorbing human insults against its environmental integrity. The desert climate and harsh temperatures preserve our accumulative abuses. In an environment where one can easily find relics of historic expeditions of centuries ago, the present waste will always stare back at us in mute testimonial to our arrogance. Items now left behind in the age of air travel can no longer be regarded as "relics", just refuse.

This survey, together with that of Benton's (1991) for Ducie Atoll, indicates that though some of the most remote corners of the globe may be technically classified as "wilderness" through absence of human inhabitation, nowhere is really free of human waste.

Table 14.1. Litter collected from Ellesmere Island.

HARVEST FROM ELLESMERE ISLAND	
Gasoline drums (45-gallon, several stencilled "Eureka")	47
Rusted metal food containers	41
Rifle cartridges	15
Gasoline drums (25-gallon)	14
Snowmobile parts	14
Self-agrandizing cairns by Army personnel	10
Food crates (wooden and metal, each containing about 50 items)	7
Recent news magazines	6
Metal cans and cylinders	6
Cigarette butts	6
Angle iron (3-m lengths used to weigh down plastic for cartographic bench mark, 1 with stamp '1975')	6
Pieces of clothing	5
Plastic garbage bags (empty)	5
Bales of wire	4
Scientific crates (some identified)	3
Snow gauges	3
Tarpaulins (plastic and cotton)	3
Tent heaters	2
Cotton and plastic cloth (15-m lengths)	2
Wooden boxes	2
Styrofoam drinking cups	2
Flare cannisters	2
Wooden tower (for radio antenna)	1
Graffiti (3-m letters for incoming helicopters spelling out sexual slang)	1
Surveying steel cable	1

Plate 13. Litter on Ellesmere Island: Empty flare canister at Lake Hazen discarded by Canadian Armed Forces (bottom-left), gasoline drums at Ella Bay from an unknown scientific research program (top-left), Bradley and England's abandoned equipment and food cache at Archer Fiord (bottom-right), animal remains and garbage from the Inuit community at nearby Grise Fiord found at Craig Harbour (top-right).

CHAPTER 15

Vestiges of the Past and Portents of a Troubling Future on Ellesmere Island: Litter Inventory

INTRODUCTION

"Wilderness"—the very word evokes images of vast expanses of wild and undisturbed nature, drawing like a magnet those seeking solace from the urbanized sprawl in which they find their lives entangled. Increasingly though, in an age of planetary domestication under the guise of that most dangerous of oxymorons, sustainable development, the task of finding that primordial haven, sequestered far from the reaching tendrils of what Edward Abbey has referred to as "syphilization", has become more difficult. Despite ostensibly searching for the elusive blue sheep of the Himalayas, one wonders if that for zoologist George Schaller this is rather the means than the end, when he states "this is what it's all about...to be able to go up into a valley, and not come on a pile of human dung...". Such misanthropic yearnings were also an objective of our expedition during the spring of 1990 when we traversed by ski, the length of Ellesmere Island: "Capped by ice and cut by glaciers, a broad barrier of pack ice hugged its coastline as if to preserve its purity...Ellesmere Island is a diverse and unspoiled High Arctic wilderness; a place where one can travel for months without seeing another person and where one can become immersed in the land, its wildlife and its history" (Dunn1990).

Ellesmere Island, lying between 76 and 84° N, is the northernmost land in North America, and with a surface area of 76,600 square miles, is the ninth largest island in the world. First viewed by (document recording) European explorers in 1616 and visited in 1854, the coast of Ellesmere Island was not completely surveyed until the 1930's. Due to its remoteness, Ellesmere Island had been little visited until the 1950's when Canada's nothernmost town, Grise Fiord, the military base at Alert, and the weather station at Eureka were established on the southern, northern and western coasts, respectfully. In 1988, the Parks Service of Environment Canada formerly established a national reserve of 37,775 square miles to (somewhat paradoxically) both protect the unique thermal oasis surrounding Lake Hazen and cater to tourism demands.

In *Arctic Dreams*, Barry Lopez has written passionately that in the High Arctic one has the overwhelming "feeling that this is the floor of creation", the land having just emerged from its protective cradle of ice. During our three month, human-powered traverse of the 1300 kilometers of Ellesmere Island, we came to realize that if it was curiosity in Paradise which was the original sin, then certainly the cavalier jettison of human garbage must have been a close second. Expedition organizer John Dunn has described the stunning natural beauty which we experienced during the Arctic Light Expedition (Dunn 1990, 1991); the present article documents the dark side of what we encountered along the way.

As a consequence of being an environmental scientist accustomed to pollution research, and indeed being funded during the Expedition to monitor the deposition and accumulation of airborne pollutants on Ellesmere Island, I had steeled myself against the temptation to view the High Arctic as "pristine". Still, I and the other members of the expedition were ill prepared for the magnitude of the garbage trail which we followed from one end of Ellesmere Island to the other, thereby completely shattering the myth of an unspoiled Canadian Arctic wilderness.

Inventory

(1) March 31; Cape Columbia/Arctic Ocean

Here on June 8, 1906, Peary, returning from his explorations along the northern coast of Ellesmere Island, ascended the 1,800-foot eastern peak of Mount Cooper Key and erected a massive cairn. Quite possibly, not until 1953 was this site visited again, at which time Peary's record and piece of his polar flag were retrieved. Today, with the existence of the military base at nearby Alert, this historic site and nearby Cape Aldrich where Peary's famous signpost stands, have been egregiously belittled with a cluster of self-aggrandizing monuments by Army personnel safely deposited at this location, the northernmost point of land in North America, from the comforts of some aircraft. The most obtrusive of the cairns from these modern day intrepid "explorers" is the one standing beside the partially dissembled remains of Peary's original atop Mount Cooper Key. Here, complete with bass relief picture, is a monument to Prince Andrew, declaring that no member of the Royal Family has ever been at this northern latitude.

(2) April 6; Head of Gypsum River on the Arthur Laing Peninsula

At the top of the divide, in an area which we felt confident from our exhaustive reading of the historical literature that no southerner had yet traversed on foot, we discover what is to become an all too frequent occurrence–two empty 45 g gasoline drums.

(3) April 26; Ruggles River/Lake Hazen

Lake Hazen, the largest lake within the Arctic Circle, was not "discovered" by southerners until 1882, and because it is a unique thermal oasis, is now the centre of North America's nothernmost protected reserve. Here, at the Ruggles River, the only outlet to the lake, Greely found extensive Palaeoeskimo ruins. And here, 108 years almost to the day after Greely first visited this marvelous location, our expedition encountered a much different historical legacy. For several days while we camped on the opposite (northern) side of the lake, Canadian Armed Forces helicopters repeatedly landed at the Ruggles River outflow to go icefishing. Later, upon visiting the site of this unsupervised fishing in a national protected reserve, we found litter scattered about the dozen or so fishing holes: discarded cigarette butts, paper cups labeled in marker "corn for icefishing" over top of the patterned insignia "Her Majesty's Canadian Armed Forces", and several empty flare canisters which had been thrown out of the helicopter to judge wind direction prior to landing. The nearby ancient Palaeoeskimo ruin described by Greely was stuffed full of a large cache of similar military debris. An Arctic fox, obviously accustomed to a diet of fishing scraps or other garbage left behind by similar military visits, purposely inspected all the fishing holes before returning to the more natural environment.

(4) May 3; Unnamed river b etween Murray Lake and Simmons Bay

Here, moments after traversing the snow free, azure ice surface of Murray Lake, truly one of the most beautiful scenes we encountered on the entire expedition, our elation is dashed by the sight of 15 (mostly full) 45 g gasoline drums which have attracted the curiosity of several Arctic hares.

(5) May 5; Ella Bay/Archer Fiord

Again, we find two empty 45 g gasoline drums. Expedition member, Mike Sharp uses his ice axe to repeatedly puncture the drums and then roll them off the delta, over the icefoot, and into the tidal crack, his philosophy being that come summer thaw they may "disappear" underwater and no longer blight the view (i.e. the dust-under-the-rug approach to pollution mitigation).

(6) May 5; Southern end of Ella Bay/Archer Fiord

Five kilometers further south on a large braided delta at the head of Ella Bay, in an area encircled by the grandeur of the towering Victoria and Albert Mountains, we encounter the largest garbage dump we are to find on our expedition. Located here just south of the Park boundary, and spread over tens of square meters, are 15, 45 g gasoline drums (some still filled), 8 smaller 25 g drums, and a huge cache of scientific paraphernalia, some of it contained within reinforced metal or wooden crates, the rest simply lying scattered about: snow gages, tent heaters, a komatik sled, bags of freeze-dried foods hermetically sealed in plastic packages of a type not seen for over a decade, a cluster of food containers, rotting boxes spilling out the rancid remains of what were once crackers, and other scientific flotsam under a weathered blue tarpaulin.

Unlike the previously encountered gasoline drums, here we have a precise identification of the culprits involved. Stenciled prominently on the sides of several of the equipment/food crates are their names. The irony is that this refuse belongs to an individual who wrote a 1983 article in which he railed against the establishment of a national park on Ellesmere Island, stating that because "Northern Ellesmere Island is an exceptional environment...it needs special protection because it is so vulnerable to long-lasting, cumulative disturbances by people". Such a Park, this individual advanced, would attract too many tourists to an "environment [which] has exceptional limits in sustaining and suffering human use, even by the most modest southern standards". If not for the horrible mess engulfing us, so incongruous against the sublime Arctic landscape, the scene is not without (black) humour. The hypocrisy of scientists who on one hand argue against allowing the public to have access to their "private" research domains because lay people, presumably lacking knowledge, can't fully appreciate the fragility of the environment, while on the other hand discarding their own scientific waste and altering the environment by their very presence, all in the name of that most noble (and perhaps therefore excusable) of pursuits, scientific research, is unfortunately not unique to high latitude regions.

(7) May 12; Head of Radmore Harbour

Once more, where there is a flat raised beach or river delta upon which an aircraft can land, the ubiquitous "flowers of the north" - 45 g gasoline drums - stand in mute testimonial to human- (usually "man") -kind's scientific exploration of the Arctic. Mike, again practicing his "out-of-sight..." mandate, punctures two drums and carries them to the tidal crack.

(8) May 24; Washington Irving Island/Dobbin Bay

Flushed with our success in locating the cache of the 1876 Nares Expedition on nearby Cape Hawks, we ascend the prominent, flat-topped summit of Washington Irving Island, a place steeped in the history of Arctic exploration. First visited by Hayes of the Kane Expedition in 1854 and next by Markham on the Nares Expedition who found two "ancient cairns, far to old to have been erected by Dr. Hayes, the only traveler known to have visited the neighbourhood", Washington Irving Island has attracted the attention of many pursuing the subject of early Norse visitation to the Canadian Arctic. By 1939 when the RCMP patrol from Craig Harbour visited the island, the monument was no longer evident. What greeted us on the summit was far from mysterious.

Here, the stones of what might have once been the possible Norse or Nares cairn had been used to weigh down long strips of now partially decomposed cotton and plastic, set out as a giant cross. The desecration was further enhanced by dozens of 3 m lengths of angle iron scattered about atop the cloth, one of which bore the stamp "1975". Obviously, we had stumbled across the remains left behind by the Canadian Surveys and Mapping Branch of the Dept. of Energy, Mines and Resources. Such crosses are set up at easily accessible places of known elevation to enable measurements of inter-distance between crosses and to calibrate elevations from the angles of aerial photographs.

(9) May 25; Cape D'Urville/Kane Basin

Here, at the overwintering site used by Captain Bob Bartlett, one of Canada's foremost polar explorers, during Peary's bid for the Pole, we find another 45 g gasoline drum labeled with "Eureka" on its side. Tracks indicate that the drum had recently attracted the attention of a wandering polar bear.

(10) June 16; Head of Makinson Inlet near Split Lake

After spending the last week high atop the Prince of Wales Icefield, some of it in harrowing gale conditions, we descend to the coast and are greeted by an Arctic spring filled with a plethora of flowers, the sounds of birds, and two 45 g gasoline drums...

(11) June 20; Makinson Inlet near Bowman Island

On a large raised deltaic beach used previously by American mountaineers and containing some marvelous Thule ruins, we find that most ubiquitous yet incomprehensible of all forms of pollution - graffiti. Scraped into the polar desert floor of broken gravel only a short distance from the archaeological site are 5 m high letters left for incoming helicopter pilots, spelling out the words "Blow Me"...

(12) June 28; Craig Harbour/Jones Sound

Just 15 km south of here, master mariner, Captain J. Bernier landed in 1907 to take possession of Ellesmere Island and all other surrounding land for Canada. In 1922, Bernier moored his ship, the *Arctic*, in the secluded bay, establishing the Craig Harbour RCMP Post which was manned for several decades. Today, the remains of this historical period are buried under the mountains of garbage deposited by the Inuit from nearby Grise Fiord who use the only building still standing as a hunting shack for themselves and their southern clients.

Feelings of euphoria and accomplishment in completing the first human-powered traverse of Ellesmere Island are soon muted as we enter Craig Harbour and see the trash signaling our reentry into Abbey's "syphilization". Dozens of gasoline drums dot the once picturesque setting. Plastic bags and recent magazines blow about in the wind between abandoned pieces of snowmobiles. Bales of partially rusted wire lie in the stream bed alongside empty cans of food. The sight of the copious flotsam is so distressing that we choose to turn our backs on Craig Harbour, our destination for the last three months and 1300 km, and instead set up camp away from the squalor.

(13) Kind Edward Point/Jones Sound

At the southernmost tip of Ellesmere Island we find a 2 m length of steel cable abandoned by a survey crew installing a trig reference point identified with a "1983" date. Therefore, from its absolute northern to southern tips, Ellesmere Island is polluted with discarded human waste.

Summary

Far from being the untainted wilderness one might expect, Ellesmere Island is riddled with the detritus of human passage and occupation. On average, we encountered about one piece of garbage every 4 km of non-glaciated terrain traveled. This is shocking and embarrassing as Ellesmere Island is by all measures still one of the most remote areas of the globe, with only about one hundred permanent residents on an island the size of Great Britain.

The High Arctic is extremely unforgiving when it comes to absorbing human insults against its environmental integrity. The desert climate and harsh temperatures preserve as a permanent record the litany of our accumulative abuses. In an environment where one can easily find relics of historic explorations of centuries ago, the present waste will always stare back at us in testimonial to our human arrogance. Items now left behind in an age when the advent of air travel has opened up the High Arctic, can no longer be regarded as "relics", just refuge.

Problems and Solutions

Gasoline Drums—No single litter problem is more pervasive in the Arctic then that of discarded 45 g gasoline drums. It needn't be this way however. In Antarctica, where transportation costs are exorbitant, the idea of carrying extra weight in the form of metal gasoline containers is unpopular. In such cases, 40 ton "fuel bladders" constructed of plastic or plastic impregnated cloth are used. These have the advantage that they can be collapsed upon emptying. There is no reason why this technology could not be adopted on a smaller scale in the Canadian Arctic. As is current practice, most of the drums we encountered were deposited after the aircraft had landed to pick up the scientists (mostly geologists and their rock samples), fueled up, and then left the large drums behind because of space constraints. Had collapsible fuel bladders been used instead, these could have simply been folded and carried away rather then being jettisoned and left to visually pollute the landscape. Presuming that the technology does exist (and if not, then why not?), Transport Canada should make an immediate and concerted effort to implement recommendations or regulations to completely phase out the use of 45 g gasoline drums in the High Arctic.

Scientific Equipment—Although it has been argued that even as late as a decade ago, scientific activities within the Canadian Arctic were conducted under a different environmental ethic, and that such an abandonment of equipment would never occur today, this does little to remove the legacy of existing scientific flotsam which is strewn across the North. The Polar Continental Shelf Project, the umbrella group which coordinates all research activities within the Arctic, must continue to rigidly enforce it's garbage in - garbage out policy. Polar Shelf should more closely monitor to see that all equipment taken into the field arrives back at Resolute when the scientists are picked up at the end of their season. Financial assistance and encouragement (such as free air travel) should be given to scientists such as a group from Trent University who purposely set about to carry back to Resolute hundreds of kg of equipment left behind by others from decades of glaciological research at a site on nearby Axel Heiberg Island.

Canadian Military—Because of "grandfather clauses", the presence of noxious, environmentally unsound activities can occur within National Parks across the country. One can only hope that with the waning of the Cold War and the consequent shrinkage of our Defense Budget, that someone in Ottawa will begin to realize that the money spent on a group of individuals to protect us from the "imminent" Red Scourge coming over the North Pole, could be better used elsewhere. Alert's days and the environmental problems associated with its presence may therefore be numbered. In the interim, Parks Canada and the Base Commander at Alert must insist on a greater environmental awareness among the pilots. Low level flights below the 300 m elevation limit, construction of tacky and completely inappropriate monuments at Cape Columbia, and the practice of icefishing at Lake Hazen, should all be stopped.

Surveying Equipment—As cartographers continue to produce increasingly more detailed maps of remote Arctic regions, their discarded materials can be expected to accumulate. Is it too much to ask for the Dept. of Energy, Mines and Resources personnel to retrieve their surveying equipment after its use? More pressing, under absolutely no circumstances should historical sites be selected as surveying benchmarks in the first place. Prior consultation with Arctic historians would guarantee the preservation of the integrity of such sites free from cartographic abuse.

Squalor at Craig Harbour—RCMP officers in Grise Fiord, while acknowledging the mess that the Craig Harbour Post was in, and understandably being quite embarrassed by it, stopped short of offering to clean it up. Considering the historical significance of this location in relation to Canada's Arctic sovereignty, it behooves both the RCMP and perhaps the Heritage Section of Parks Canada to undertake to bring Craig Harbour to a state which the country can be proud of. Given the proximity of Grise Fiord and number of tourists who pass through each summer, the possibility of an RCMP historical-interpretive centre at Craig Harbour may be an idea worth pursuing. This is of course dependent on curtailing the littering activity by Grise Fiord Inuit who currently use the site as a hunting base.

Graffiti—As old as the visits of Romans to the Giza pyramids or English poets to Greek ruins, graffiti, a powerful voice of social unrest in cities, is completely out of place in natural environments. The solution to this problem is beyond fathoming.

Waste and Garbage from Adventure Expeditions—During our expedition we burned our toilet paper and saved all garbage to have it either flown out by incoming scientific groups at our two cache sites or to carry it with us to the end. It is therefore completely feasible for a lightweight expedition to spend almost a hundred days traveling in the High Arctic and leave behind little record of their passage over the landscape. Contrast this with other expeditions going to the North Pole which read like a diary of littering in the act. Ward Hunt Island near Cape Columbia where we began our trip, and which is the traditional debarking site for North Pole expeditions, is strewn with the garbage of many famous modern day adventurers. It would be hypocritical in this article not to point a finger also at fellow adventure travelers who have themselves mistreated the precious and beautiful environment through which they traveled.

Plate 14. Litter on Ellesmere Island, part II: Dissembled historic cairn and recent angle iron used to weigh down plastic sheets at a cartographic survey site on the summit of Washington Irving Island; expedition member M. Sharp carrying a punctured fuel drum over the icefoot to the tidal crack in the Kane Basin.

CHAPTER 16

Trash Talking: In Defense of Litter Surveillance

WHAT PRICE ARCTIC RESEARCH?

Professor John England has responded to a column by Stephen Strauss [see Appendix III] concerning and article of mine which recently was appeared in the science journal *Nature*. In his April 11 letter [see Appendix III], Dr. England makes several unfortunate factual errors and leaves some profound misconceptions with the public on issues of funding and responsibility for research, as well as how humans perceive environmental pollution.

The article in *Nature* which has spawned the present controversy was a straightforward and factual listing of all the human debris we encountered during a 1990 ski-traverse of Ellesmere Island. My "moralizing" as Dr. England refers to it, consisted of the following statement: "The high Arctic is extremely unforgiving when it comes to absorbing human insults against its environmental integrity. The desert climate and harsh temperatures preserve our accumulative abuses. In an environment when one can easily find relics of historic expeditions of centuries ago, the present waste will always stare back at us in mute testimonial to our arrogance. Items now left behind in the age of air travel can no longer be regarded as relics, just refuse…Though some of the most remote corners of the globe may be technically classified as 'wilderness' through absence of human inhabitation, nowhere is the world really free of human waste."

Far from "finger-pointing", in the *Nature* article I neither mentioned Dr. England's name nor identified the sexual slang permanently scraped into the desert floor next to a historical site (not merely "trampled in the snow" as Mr. Strauss incorrectly states). Dr. England's vitriolic reference to my naiveté, inanity, inability to 'realize', and perpetration of written "garbage" (!), is misdirected and goes far beyond the bounds of my simple article in *Nature*. In reality, it is the journalistic interpretations of my findings by Mr. Strauss which have so angered Dr. England. During my interview with Mr. Strauss I refused to identify Dr. England. However, it is the job of reporters, especially ones such as Mr. Strauss who have reputations for being scientific watchdogs, to investigate backgrounds to their columns and to uncover and mention names if they believe it furthers their cause, in this case, obviously one, in Strauss's mind, of egregious hypocrisy.

There is no doubt that Dr. England comes off badly in the column by Mr. Strauss, thereby explaining the insulting tone of his response. In a way this is unfortunate, as Dr.

England was by no means the sole perpetrator behind the discarded garbage trail we discovered on Ellesmere Island. Other litter belonged to the Canadian Armed Forces from Alert, cartographers from Ottawa, Inuit from Grise Fiord, and adventure tourists from a variety of countries. Dr. England merely had the double distinction of having his name stenciled on the crates left behind, as well as writing a feature article in *Canadian Geographic* in which he went to great length about the exceptional sensitivity of northern Ellesmere Island to such human degradation.

Despite what Dr. England states, neither my original article nor Mr. Strauss's "opportunistic journalism" "trivialize" the work of arctic scientists. What we do, rather, is to question the manner under which the scientists have conducted their research. At what stage does the justification of that always sacred scientific cow "understanding [the] environment" merit alteration of that self-same environment during the process?

Underlying the entire discussion is the perception about how serious one considers to be the problem of discarded or "temporarily" abandoned (in England's case, "temporarily" for over a decade) human jetsam in the Arctic. In this respect, there is a logical inconsistency in Dr. England's letter when he states that "The column [Strauss's] also gives the impression that Arctic scientists have an arrogant and cavalier attitude toward the environment", and then continues two sentences later: "If Dr. France would like to make a career out of playing Sherlock Holmes, I would gladly point him in the right direction! On numerous occasions we have picked up debris left behind by others." This was exactly the point that Mr. Strauss and I were making in our independent articles. The fact that R. Wissink of Parks Canada stated (to Mr. Strauss) that his organization has had to carry "away 25,000 kg of refuse from [Arctic] research sites" established during a single year, highlights the magnitude of the problem. Again Dr. England seems confused, dismissing the importance of finding litter as a non-issue on one hand, yet later referring to his own abandoned garbage as "a problem". Or is he simply referring to two different things here, the only real "problem" being to his reputation and not to the environment?

Dr. England states that "Because the remainder of [his] cache was removed over a year ago (solving the problem) the reader can only conclude that Dr. France's finger-pointing is inane. He is beating a dead horse". In fact, I was heartened to discover only two weeks ago through Mr. Strauss's research, that the cache had been "cleaned up by Parks Canada and charged against his [Dr. England's] research funding." Heartened, because it is reassuring to environmentalists to know that individuals can possibly make a difference. Perhaps this unilateral cleanup by Parks Canada was a result of the two hour conversation I had with them upon returning from the 1990 expedition, or perhaps it was due to the national coverage given to the topic by John Dunn, leader of our expedition, whom had been traveling back and forth across the country giving a popular speaking tour funded by the Royal Canadian Geographic Society, in which he dramatically and repeatedly showed slides of the material left behind by Dr. England (among others).

But how really serious is the "problem" of visual marring of the Arctic landscape? Dr.

England, although he states that he does not "condone it", nevertheless trivializes our finding "of one piece of litter every 4 km of non-glaciated surface traveled." He suggests that if I want to really alert the public about what Arctic pollution I should "rather concentrate on other issues such as the toxic "Arctic haze" that invades the atmosphere each winter."

My co-expedition members and I did not purposely set out to make an inventory of the litter trail which we followed from one extreme tip of Ellesmere Island to the other, thereby completely shattering the myth of an unspoiled Canadian Arctic "wilderness". We were just so appalled by the extent of the discarded material we came across during our journey that I felt the public should be made aware of the extent of the condition. In fact, I was specifically funded by the World Wildlife Fund to investigate the deposition and accumulation of long-range trace contaminants such as cesium, lead, mercury, and PCBs. This work is presently being prepared for publication in collaboration with some of the country's most respected environmental and arctic scientists. I trust Dr. England will find this research useful and thank him for highlighting its importance to a national audience.

Where I disagree with Dr. England is in his belief that visual litter is not something about which the public need overly concern themselves. This is untrue, for the simple reason that it is often the *perception* about an environment that becomes the central and most heatedly debated issue in environmental protection. The international call to halt the east coast seal fur trade the last decade was not based upon scientific arguments of seal population sustainability nor upon estimates of stock recruitment of fish prey, but rather upon the visual perception left by the harsh images of baby animals being clubbed to death. Present-day opposition to clearcut logging is seldom founded upon discussions of how timber harvesting alters biogeochemical cycles or sediment transport to fish spawning beds, but rather upon images of massive tracks of denuded forests often accompanied by a Shakespearean quote about land ravishing. These visual perceptions by the public concerning environmental degradation are so deep-rooted that the B.C. Ministry of Forestry actually advises logging companies about how to conduct harvesting so as to hide it from avenues of public transit as much as possible. Visual perceptions of what constitutes "pollution" therefore cannot be simply dismissed.

The Arctic has always attracted southerners (explorers, tourists, and scientists) through its hauntingly sublime vistas where one can seek solace from the urbanized sprawl in which lives become entangled. "This is what it is all about" wrote zoologist George Schaller who accompanied naturalist Peter Matthiessen to the Himalayas in search of blue sheep and snow leopards, "...to be able to go up into a valley, and not come upon a pile of human dung...". Even if most will probably never be able to afford the luxury of visiting the High Arctic, just to hope that perhaps there might be somewhere left on the globe that is free of human waste is reassurance as we continue to abuse our immediate local surroundings.

Scientists like Dr. England and myself, because our transportation and research costs are largely supported by the public, are fortunate to be able to experience the magic of visiting such special and fragile Arctic regions. It behooves us to be especially careful about our activities there least we be accused of being hypocritical, particularly if we take strong stances against allowing influxes of tourists into the same area. In the end, I think Dr. England, despite what he recently wrote in his overemotional (i.e. the obviously facile and politically belittling "So, in the midst of our constitutional wrangling it is appropriate that England finally respond to France.") and mean-spirited letter, agrees with this stance. For in 1983 he wrote convincingly about the "permanent" and "accumulative" "scars" which remain behind from the very research camp that Mr. Wissink of Parks Canada called "a national disgrace" in the column by Mr. Strauss which raised Dr. England's dander when it dealt with him personally (it is interesting that Mr. Wissink, representing an organization that Dr. England has had to and no doubt will continue to deal with professionally, comes off completely unscathed by Dr. England's poisoned pen).

Dr. England and I do differ, however, in our interpretation of the seriousness of this problem. He concludes that I "was in paradise but simply failed to realize it!" Not true. I, and the other members of my expedition, merely have a more exacting and discriminating definition of what should be and what should not be in that earthly garden of our imaginings.

Although much discussed in both Mr. Strauss's column and Dr. England's letter, the argument that because research costs are so exorbitant in the Arctic, that somehow this condones a wavering in vigilance about what is left behind, is weak. This is the defense, eschewing personal responsibility, that is unfortunately all too characteristic of the shortsighted manner in which many industries conduct their business; i.e. not factoring in the inherent costs necessary for cleaning up the mess after finishing and moving elsewhere. If my article in *Nature*, Mr. Strauss's column, and the present exchange between Dr. England and myself, has any lasting merit, it is in bringing to the fore the grand illusion under which arctic research has hitherto been conducted: namely, the fallacy of what the true costs really are.

Ultimately, it is the taxpayer who has to pay the price of cleaning up both the antecedent litter as well as the undertaking of more environmentally defensible (and therefore costly) research in the future. As to specifically what sector of society this should be funded by, is debatable. For example, should the burden of cleanup be undertaken solely by (a no doubt already financially strapped) Parks Canada, or should costs to remove adandoned equipment be automatically added onto every research budget funded by the National Science and Engineering Research Council? Should we as taxpayers insist on an increased vigilance by the Polar Continental Shelf Project, the umbrella government organization that coordinates research in the Canadian Arctic, to better enforce an "equipment/garbage in and out" policy to all scientists relying upon its support?

Dr. England is only partially correct in using the all too convenient scapegoat of "modest funding" under the auspices of which arctic research is customarily conducted. Environmentalists would argue, however, that in this case, the ends may not justify the means, and that perhaps until the Federal Government (and us as taxpayers) seriously fund arctic research to its true cost, that we should exercise extreme humility not hubris when we attempt to argue as does Dr. England that "Canada is a world leader in understanding

Arctic environments". We must seriously ask ourselves at what price has this been accomplished? Has our relentless search for Baconian knowledge come at the expense of our loss of better Gaian wisdom? Only recently we learnt the sober fact in the State of the Environment Report that Canada is also the world's leader in the per capita production of waste. Not really much to be proud of in the end.

Dr. England seems to blame others for the deterioration of his 15-year old, uncollected food and equipment cache. However, in the city, if one leaves out garbage cans on the edge of the street which, perhaps because of a labour strike are left uncollected for several days and are tipped over by vandals, then it is the responsibility of the originator of that material to have it cleaned up, not the City.

Although, the "problem" may have been "solved" from Dr. England's point of view with removal of his cache of equipment, thereby soothing his own conscious and wounded pride, the much larger problem about how we all conduct our day-to-day activities, research or otherwise, is far from being so easily settled. Were that it were so. Unfortunately, increasingly these days, the work of environmentalists seems ever more frustratingly to resemble the eternal labours of Sisyphus. Progress, if measured at all, comes only incrementally so.

Paradise Lost?

SIR—Dr. England has responded to a *Nature* article of mine concerning garbage on not-so-pristine Ellesmere Island in the high Arctic, which has since publication been covered in over half a dozen newspapers or magazines around the world. Unfortunately, England makes several factual errors and leaves some profound misconceptions with *Nature's* readership on issues of responsibility for research, as well as with how humans perceive environmental pollution.

England justifies abandoning his scientific equipment cache because of "limited financial support". This is also the oft-heard defense that is characteristic of the shortsighted manner in which many industries conduct their business; i.e. not factoring the inherent costs necessary to clean up after the work is done. If this exchange has any lasting merit, it is to seriously question the illusion that arctic research always covers its entire, life-cycle costs.

England says because his cache was "removed" to Resolute Bay thereby "solving the problem", he fails to see the point of my argument. He neglects to mention, however, that because his cache had long been known as a major eyesore to Parks Canada officials through numerous complaints fueled partially by our 1990 expedition, his "mess...has in fact been cleaned up by Parks Canada and charged against his research funding" (my italics). No wonder that England has admitted to being somewhat "red-faced" about the whole incident.

England says that a previous article which he wrote "never implied that tourists should be excluded [from northern Ellesmere Island] while scientists enjoyed the area as a private sanctuary." England is correct here in that, true, he did not "imply" any such thing; indeed, he seemed to explicitly state it. The subtitle of this article by him (England 1983) is "A case against a national park at the top of the world". Here he correctly states that "on northern

Ellesmere Island, despite the best intentions, there remains an inherent conflict between parks for people and parks to preserve wilderness", and that "this proposed tourism clearly contradicts the objective of keeping such a park 'unimpaired' for future generations". He then argues that "our most exceptional, sensitive areas (such as around Lake Hazen) may be better set aside for non-use." Yet England, who has "worked on northern Ellesmere Island for the past 11 years", left his own extensive cache only a few kilometers away from this now established National Park Reserve.

England seems confused about what does and does not constitute abandoned waste. Elsewhere (England 1983) he has written emotionally about the "permanent" and "accumulative" "scars" which remain behind at a research camp from which Parks Canada has since had to carry "away 25,000 kg of [scientific] refuse" which was "a national disgrace". Yet England becomes hypocritically upset when he is identified (not by me) by the press for likewise leaving behind equipment and decomposing food contained within nearly two-decade old and damaged foil bags, broken tins, and rotten boxes. I for one would not have wanted to rely upon the latter for "sustenance" to use England's word for the remains.

England states that the Polar Continental Shelf Project has had a "garbage in garbage out" policy for more than a decade, failing to mention (a) that its instigation came from a bottom-up protest by a few independent scientists and numerous environmentalists, not through a top-down bureaucratic decree (J. Svoboda, pers. comm.); and (b) that as I stated, it is its enforcement, not its conception, that is the real issue at the present time.

There is a gross inconsistency in England's writings concerning the magnitude of abandoned and visually marring litter left behind by scientists in the Arctic. In *Nature* he stated that "most field work in the Arctic is conducted by conscientious researchers who cherish this environment." Yet less than two months ago he wrote that he could easily point out many instances of the cavalier jettison of scientific refuse, indeed even going as far to state that "on numerous occasions we [him and his graduate students] have picked up the debris left by others." He appears to be very confused. Scientists, it seems, if we are to believe the earlier statements by England, are by no means free of guilt when it comes to discarding "temporarily" unneeded or conveniently "forgotten" material.

England's tangential discussion about glaciation history and biological refugia represents the sort of classic (and in this case, humorous) attempt to change the subject at the same time as enabling him to cite a handful of his own publications in order to demonstrate his scientific credentials, something which has never been questioned.

England trivializes my finding of "one piece of litter every 4 km of non-glaciated surface traveled" on Ellesmere Island. I maintain that these findings, together with those of Benton (1989) for an equally remote location, are both serious and embarrassing because they underscore the lack of respect toward the environment under which we humans conduct our daily lives, research or otherwise.

Finally, England concludes that I "was in paradise but simply failed to realize it". Not true. I and the other members of the Ellesmere Island ski-expedition merely have a much more exacting and discriminating definition of what should be part of our Arcadian imaginings. Sadly, Lopez's "floor of creation" (1987) is beginning to resemble Milton's "Paradise Lost" in terms of its accumulated litter, something that is bothersome to all but the most misguided despoilers.

APPENDIX I

Location, Temperature, and Travel Distance Log

Date, location, average temperature (°C) and distance traveled (km) during the 1990 Trans-Ellesmere Island Ski Expedition. Maps (1:250,000) from north to south are: Clements Markham Inlet 120 F & 120 G, Lady Franklin Bay 120 C & 120 D, Kennedy Channel 120 B, Dobbin Bay 39 H & 29 G, Ekblaw Glacier 39 F & 39 E, Sawyer Bay 39 G, Strathcona Fiord 49 E, Vendom Fiord 49 D, and Craig Harbour 49 A.

March 30, 31	Cape Aldrich/Columbia	-25	2, 15
April 1, 2	Parr Bay	-27	9, 8
April 3-8	Gypsum Creek	-31	6, 6, 10, 11, 11, 15
April 9	Clements Markhm Inlet	-18	9
April 10-17	Piper Pass	-18	9, 13, 0, 7, 5, 6, 16, 17
April 18-20	Hazen Trough	-28	15, 13, 12
April 21, 22	Lake Hazen	-24	16, 0
April 23, 24	Cuesta Creek	-23	13, 10
April 25, 26	Lake Hazen	-22	9, 2
April 27	Ruggles River	-17	15
April 28, 29	Chandler Fiord	-15	15, 16
April 30	Ida Bay	-13	10
May 1, 2	Dodge River	-12	14, 18
May 3	Murray Lakes	-8	17

May 4, 5	Archer Fiord	-4	8, 16
May 6	Christiansen Lake	-11	16
May 7-12	Jolliffe Glacier	-9	5, 10, 8, 12 17, 16
May 13	Radmore Harbour	-3	20
May 14	Rawlings Bay	-2	8
May 15	Cape Joseph Good/Wilkes	-4	13
May 16	John Richardson Bay	-4	15
May 17	Cape Collinson/Joiner Bay	-4	14
May 18	Scoresby Bay	-6	23
May 19, 20	Darling Peninsula	-5	16, 22
May 21-23	Dobbin Bay	-5	6, 0, 18
May 24	Cape Hawks	-10	14
May 25	Allman Bay/Cape D'Urville	-9	15
May 26	Franklin Pierce Bay	-10	19
May 27	Princess Marie Bay	-1	28
May 28	Bache Peninsula	-2	21
May 29, 30	Flagler Bay Polynya	-2	5, 5
May 31	Buchanan Bay	0	17
June 1, 2	Alexandra Fiord	0	13, 18
June 3, 4	Hayes Fiord	-1	23, 10
June 5, 6	Jokel Fiord	+1	23, 13
June 7, 8	Stygge Glacier	-2	18, 17
June 9-14	Prince of Wales Icefield	-2	23, 14, 15, 12, 24, 31
June 15	Split Lake Glacier	+5	12

June 16	Makinson Head Oasis	+4	25
June 17, 18	Makinson Polynya	+5	20, 6
June 19-21	Makinson Inlet	+8	23, 14, 20
June 22	Bentham Fiord	+6	23
June 23-25	Manson Icefield	+6	25, 40, 25
June 26	Jakeman Glacier	+7	27
June 27, 28	Jones Sound	+8	15, 20
June 29	King Edward Point	+7	10
June 30	Craig Harbour/North Water Edge	+6	0
July 1	Cone Island/North Water Edge	+6	16
July 2	Jones Sound	+5	32
July 3	Lee Point/Grise Fiord	+6	5

APPENDIX II

SCIENTIFIC ABSTRACTS

CHAPTER 1

During the first ski traverse of Ellesmere Island in spring 1990, purple saxifrage (*Saxifraga oppositofolia*) was collected at 10 sites from 76 to 82° N. Measured cesium-137 levels in this cushion plant displayed a progressive decrease in activity north of 78° N, reflecting past global patterns of radionuclide fallout. Lower ^{137}Cs activity at the southern end of Ellesmere Island may reflect a northward shift in the distribution maximum since a previous latitudinal survey conducted in 1979-1980. Levels of ^{137}Cs in three species of lichens were consistently higher than those for nearby saxifrage, possibly owing to the larger exposure to fallout for much of the year and the slower rate of lichen growth. In support of previous research, no ^{134}Cs was detected, which indicated that Chernobyl fallout had not been deposited in significant quantities at these extreme northern latitudes. Specific activities in 1990 of saxifrage samples were compared with similar samples collected during 1979-1980 to derive an effective half-life of 6.2 ± yr for northern Ellesmere Island.

CHAPTER 2

Although organochlorines (OCs) have frequently been detected in atmospheric samples from the High Arctic (> 75°N), there is no evidence of their accumulation in terrestrial biota from these extreme latitudes. The present study confirms that the emission of OCs to the trophoshere from southern sources is substantial enough for these contaminants to be transported to and be accumulated in lichens (*Umbilicaria muchlenbergii*) from Ellesmere Island in the eastern Canadian High Arctic. Toxaphene was found to be the most abundant OC measured in these lichen samples (17-24 ng/g dry wt), representing about 65% of total plant burdens. The OCs a-HCH, HCB, g-HCH, s-PCB, DDE and DDT in lichens from Ellesmere Island were lower in concentration than those for lichens from Europe and southern Canada, but similar to residues measured in lichens from other remote polar sites (Spitsbergen and Antarctica). HCB and s-CHLOR concentrations in Ellesmere Island lichens were proportionally higher than those found for more southern growing species. The presence of OC residues in lichens from the High Arctic raises the possibility that browsing caribou and their wolf or human predators might also be accumulating these contaminants in detectable amounts.

CHAPTER 3

Previous atmospheric and precipitations sampling on Ellesmere island in the Canadian

High Artic (76-82° N) has detected the presence of trace contaminants originating from both mid-latitudinal and transpolar sources. Plant samples (*Saxifraga oppositofolia*) were collected along a latitudinal transect from 7 remote sites on Ellesmere Island in order to determine the most likely influx pathway of organochlorine (OC) pollution to this region. Toxaphene, HCH, CBz, CHL, and DDT residues and selected PCB homolog groups decreased with northward distance indicating the predominance of North American sources. Further, all OC concentrations in saxifrage were found to be related to plant cesium-137 activities and therefore differences in the deposition of southerly-originating pollutants. The close association between residues of these micropollutants raises the possibility that the logistically much easier measurements of ^{137}Cs uptake could be used as a convenient environmental monitor of regional patterns in OC deposition in locations where pathways of atmospheric transport are similar.

CHAPTER 5

Measuring deposition of anthropogenic lead is essential for quantifying global patterns of pollution dispersion. Concentrations of Pb, however, have only rarely been measured in lichens collected from the High Arctic. In this study we used flameless atomic absorption spectrophotometry to determine that the average Pb content in 5 lichen species from several sites on central and southern Ellesmere Island was 2 $\mu g\ g^{-1}$ dry wt. These are among the lowest Pb concentrations that have ever been recorded in Arctic or any plants. Literature data were combined to establish a latitudinal gradient showing decreasing Pb concentrations in lichens with increasing latitude throughout Canada.

CHAPTER 6

Previous atmospheric and precipitation sampling on Ellesmere Island in the Canadian High Arctic (76 to 83° N) detected the presence of trace contaminants originating from both mid-latitudinal and transpolar sources. Plant samples (*Saxifraga oppositofolia*) were collected along a latitudinal transect from 10 remote sites on Ellesmere Island in order to determine the influx pathway of lead pollution to this region. Unlike contemporaneous measurements of cesium-137 activities and organochlorine concentrations, the bioaccumulation of Pb was greater in northern compared to southern Ellesmere Island. Stable isotopic measurements of Pb suggested that Eurasian sources contributed over half of the Pb burden in the most northern-sampled plants. These data support stable isotopic measurements of aerosols from northern Ellesmere Island in suggesting that Eurasian sources of Pb pollution may now predominate in this region of the North American High Arctic.

Chapter 7

In late winter 1990, the first surface transect of wildlife was undertaken through North America's most northerly protected reserve. Located in this region on Ellesmere Island is the unique thermal oasis surrounding Lake Hazen, the largest lake within the Arctic Circle. Abundance of non-migratory wildlife was considerably greater inside the Lake Hazen Trough compared to the depauperate desert landscape existing both north or south of the oasis, and was correlated with geographic proximity to the locus of Lake Hazen itself. The Lake Hazen Trough is therefore an ecosystemic centre of organization whose vegetative productivity, although generated during the brief summer period, is still sufficient to support elevated levels of wildlife abundance throughout the entire year.

Chapter 11

Seven colonies of Ivory gulls comprising 330 adults were discovered on nunataks emerging from the Manson Inlet Icefield of southeastern Ellesmere Island. This brings to total 33 the number of active breeding colonies that have been located in North American over the 14 yr since the first was discovered.

Chapter 12

A Rough-legged Hawk (*Buteo lagopus*) was observed on two separate occasions over a small localized area at the head of Makinson Inlet on southeastern Ellesmere Island. This is the most northerly record for this species and a first for Ellesmere Island.

APPENDIX III

News Articles

Before, during and after the expedition, a series of articles appeared in *The Calgary Herald*, the hometown newspaper of expedition conceiver and organizer John Dunn. The major post-expedition article, written by John Dunn and accompanied by a sampling of his spectacular photographs, appeared in an issue of the *Canadian Geographic*, one of the major sponsors of the expedition. Another article, one whose text expedition member Mike Sharp thought did a much better job at capturing the spirit of our outing albeit with photos not of the same caliber of those of Dunn's, was written by Graeme Magor and appeared in an issue of *Hamilton This Month*, a glossy magazine from his hometown.

The following articles are those with a focus on me and mention some of the environmental research I undertook during the expedition. Readers must realize that that in many news articles, journalists tend to sensationalize certain things that they personally fixate on during the long interview process as well as to try to personalize general statements made by their subject in quotation marks even if those statements had originally been ascribed to others. Therefore treat the following articles with a healthy degree of caution in regard to what actually transpired during the interviews.

Team to Tackle Rugged North — K. Rollason, *The Winnipeg Free Press*, December 27, 1989

Cross-country skiing and camping for 100 days in temperatures averaging –30°C might seem crazy to anybody but a Winnipeger.

But that's what Bob France, 32, plans to do during a 1,400-kilometre crossing of Ellesmere Island this spring as part of the Arctic Light 1990 Trans-Ellesmere Island Expedition.

"I'm the neophyte in the group, because I wasn't drafted until a few weeks ago," says France. "The furthest north I've been before this is The Pas and Lynn Lake."

The expedition's goal is to make the first complete ski traverse of Ellesmere Island, which is as little as 765 kilometres from the North Pole.

France, who now lives in Toronto, says the trip will give him a personal high: It will make him one of the few people who have ever stood at both the northernmost and southernmost points in Canada.

Shortly before he leaves, he plans to drive to the southernmost point in Point Pelee National Park in Ontario.

"Everybody looks at maps and sees that Canada has a southern border but it seems to stretch out into the north forever. This trip will increase awareness that Canada isn't infinite," says France, who's in town visiting his parents over the holidays.

France, an ecologist specializing in the effects of lake acidification and global climate change, is the fourth member of a team that will ski the length of Ellesmere Island, from Cape Columbia, Canada's northernmost settlement.

During the expedition, which begins March 24 and lasts until the end of June, each team member will pull a 90-kilogram sled laden with supplies through days with 24 hours of sunlight.

They will ski past Lake Hazen, the largest freshwater lake above the Arctic Circle, and climb Mt. Barbeau, at 2,616 metres the highest mountain in Eastern North America.

Other members of the team are team leader John Dunn, 35, of Calgary, Graeme Magor, 33, of Hamilton, and Mike Sharp, 38, of Vancouver.

The estimated $20,000 cost of the trip will be offset by corporate donations and by preselling a limited-edition photograph of the trip.

Enroute, the team will measure the amount of ultraviolet sunlight received there to help determine how much the Earth's ozone layer has broken down.

France, who also specializes in the historical and spiritual basis of human perceptions about the wilderness, expects to write a book on his experiences during the trip. He also hopes the team will find evidence of two Arctic expeditions launched during 19th century as well as RCMP outposts established in the 1920s.

In 1881, American Adolphus Greeley lost 19 of his 26 men during a trip similar to the one France is on. Seventeen years later, Norwegian Otto Sverdrup led 20 men to Ellesmere Island for a mapping expedition and returned four years later having only lost two men.

But France isn't worried about repeating the fates of those early explorers in the sub-zero temperatures: He has skied the back country of the Rockies several times.

"Compared to the back country of the Rockies, where you constantly have to dodge avalanches, this won't be that bad. Polar bears are always a pressing concern in this area, but we've been given permission to carry rifles and pop guns."

The team will also have winter and survival gear earlier expeditions could only dream of, including a radio and a satellite location beacon.

More of concern to the team is how the four will react to 100 days of solitude in the Canadian Arctic. France says an added wrinkle is that he's never met two of the trip members and that he will have to survive three months with them or die.

This will be France's first trip to the Arctic, but the other members of the team are old hands at going to areas with sub-zero temperatures. Each has gone on expeditions to the Arctic or Antarctic.

Sharp is currently part of the 1990 Trans-Antarctica crossing of the continent which reached the South Pole on Dec. 11.

France hopes the expedition will increase awareness of the Arctic as something more than just a vast resource mine.

"We want to show the wilderness is not a wasteland. The Arctic is a place of extreme beauty and environmental sensitivity."

Going for Your Dreams - in the Arctic — J. Picton, *The Toronto Star*, March 15, 1990

Even as you read this, Robert France is standing on the most northerly point of Canada and committing a symbolic act.

It's an act that may well confuse geologists in years to come.

He is tipping sand he collected recently from Canada's most southerly point, Point Pelee on Lake Erie.

Later this year, after the adventure of a lifetime, he will return to Lake Erie and sprinkle soil he collected in the North.

"This will be, in a sense, my 'last spike,' inter-linking this vast and special country I call home," says France, referring to the building of the railway that linked eastern and western Canada.

But first, there's the matter of that 1,200-kilometre (745-mile) trek down Ellesmere Island, through valleys and past frozen fiords that have been locked in place by ice for eons.

"This will be the longest human-powered trip in the Canadian arctic since the mid-19th century," says France, a 32-year-old environmental scientist from the University of Toronto.

"Our route will take us past Lake Hazen, the largest lake above the Arctic Circle, and over Mount Barbeau, the highest land in North America east of the Rockies."

It's part of what team member John Dunn calls "going for your dreams" in lectures at schools in his locality.

His message: "Don't sit around when you're 50 wishing you'd done it when you were younger."

With Graeme Magor, a physician from Hamilton, and Mike Sharp, a Vancouver-based mountain guide, France will be part of a group called the Arctic Light 1990 expedition.

One reason for their three-month trek is to collect the most northerly samples of terrain ever taken, and to document the long-range fallout effects of organic pesticides, industrial mercury, lead emissions from cars and concentrations of sulphur from power plants.

Pollution in North Irks Ecologist — M. Zeilig, *The Winnipeg Free Press*, July 28, 1990

Ellesmere Island, Canada's northern most land mass, is starting to show signs of pollution, according to a leading ecologist.

Dr. Robert France says the island is littered with 45-gallon fuel drums, left behind by both government and private industry.

As well, historic rock cairns, some built by early Arctic explorers, have been partially destroyed, France said in a recent interview.

The former Winnipeg native said the federal government must enforce a strict anti-pollution policy in the North.

Organizations operating in the fragile environment should be forced to remove all garbage they generate in the area, he said.

France encountered the debris earlier this year during the Arctic Light Trans-Ellesmere Island cross-country ski expedition.

"The Canadian government must enforce a strict Arctic policy of garbage in, garbage out," he said.

During the odyssey, France collected the globe's northern most samples of a ubiquitous lichen known as purple saxifrage.

"I was sampling the plants for a bio-monitoring study," said France, a University of Toronto ecologist specializing in the effects of lake acidification and global climate change. He is also working closely with the Winnipeg-based Fresh Water Institute.

"The land is not pristine. We will be looking for long-range trace elements such as lead, mercury and pesticides. We're interested in the cycling of trace contaminants in the food web. And we'll also be analyzing for radioactive fallout, which is a post Chernobyl effect."

France also used a special instrument to make checks on the sun's ultra-violet light. A large hole in the Earth's ozone layer above the Arctic allows potentially dangerous ultra violet levels into the atmosphere.

"I also did the first major wildlife survey of Ellesmere Island from the ground," France said. "Areas called polynyas, which are bodies of water kept permanently open by ocean currents, serve as biological hot spots. They attract large congregations of wildlife. We saw a plethora of geese, ducks and other birdlife, seals, walrus and polar bears. Thankfully we didn't have any close calls with the latter."

"Polynyas have an effect larger than their boundaries. They are like a desert oasis, and very fragile."

France said he is concerned that some people may want to transport oil through the polynyas. He stresses that more studies have to be done on them.

Beginning on March 30 from Cape Aldrich on the Arctic Ocean and concluding 96 exhausting days later, on July 5, at Grise Fiord (on Jones Sound), the expedition travelled 1,300 twisting kilometres north to south.

"This was the first human-powered crossing of Ellesmere from tip to tip. It had never been done before," he said. "I went because I had never been up to the Arctic before. I jumped at the opportunity to join the expedition when a friend, who was supposed to go, declined, and suddenly a slot opened up."

Each participant pulled a sledge with 90-100 kilograms of provisions. Two caches of food were made earlier at strategic locations during the plane trip up to Cape Aldrich.

"It was arduous hauling the sledges. But it would have taken a lot of time to become familiarized with dog teams," said the 32-year-old France, who is spending the summer at his parent's Charleswood home. "We wanted to travel slowly and become immersed in the landscape."

The expedition was organized by Calgary resident John Dunn, a 35-year-old British-born photographer, writer, geologist, and arctic adventurer. Team physician Dr. Graeme Magor, 33, and veteran mountaineer and Antarctic explorer Mike Sharp, 38, completed the quartet.

High Arctic Trek — J. Stirling, *University of Toronto Bulletin,* October 9, 1990

Here in Toronto—with more than its share of fast-food outlets—it is difficult to imagine how Robert France felt last spring as he stood atop a polar ice cap in the high Arctic dreaming of junk food. But France remembers his cravings well. After a 96-day trek across the frozen landscape of Ellesmere Island, he and three fellow travellers couldn't wait to reach Calgary to eat ice cream and candy. "I had developed a real sweet tooth for chocolate bars and would have done just about anything to get one on the trip," he says.

France, a research associate with the University's Institute for Environmental Studies, was part of the Arctic Light 1990 Trans-Ellesmere Island Expedition, a 1,300-kilometre journey from Cape Columbia in the north to Grise Fiord in the south, that began on March 24 and ended July 5. John Dunn, a geologist and photographer from Calgary, organized the trip. The other members were Michael Sharp, who leads expeditions to the Antarctic, and Graeme Magor, a doctor from Hamilton.

France is an environmental scientist specializing in the study of lake acidification and changes in global climate. He is part of a team investigating the effects of airborne pollutants on Arctic ecosystems. There he collected samples of purple saxifrage and lichen. Starting next summer the samples will be tested for organic pesticides, mercury, lead, sulphur and radioactive fallout from Chernoboyl. "None of the compounds we're looking for is used within 800 miles of the island," France says. The study, supported by the World Wildlife Fund, includes government scientists and Professors Josef Svoboda, a Erindale College biologist, and Pamela Stokes of Trent University, former director of the Institute for Environmental Studies.

France and his colleagues were the first to cross Ellesmere Island without snowmobiles or dogs—the longest such trip in the Arctic since the 19th century when British explorer John Franklin went in search of the Northwest Passage. For 10 hours a day, they skied cross-country pulling sled-loads of 200 pounds containing food, sleeping gear and other materials. When the snow became too deep, they left their provisions, skied half a kilometre ahead to make a passable trail, then returned for their sleds.

His first glimpse of the Arctic Ocean was breathtaking. "We climbed a mountain for two hours, crested, and saw open water—icebergs and Greenland 40 miles away." Intense periods of association with the landscape were offset by periods of dissociation when life "just crawled" in front of him.

Along the way, the group conducted the first land-based wildlife survey of Ellesmere Island, including surveys of muskox and polar bears. While the desert landscape afforded relatively few encounters with animals and birds, the group did see a variety at polynyas—bodies of water kept open by ocean currents. Ducks, geese, seals and walruses gathered at the oasis-like watering holes. The explorers also found Inuit ruins—tent rings, food caches and animal traps, some dating to the 17th century. "It was humbling to see the remains of their camps and then go back to our down sleeping bags, nylon tents and gas stoves." They also came across cairns left to mark the paths of European explorers more than a century ago.

The temperature dropped to minus 35 degrees Celsius at night and despite their down-filled sleeping bags the men woke with numbed feet—a condition they lived with constantly for the first month. "It took two hours every morning just to get some feeling in our toes and even during the day our feet didn't lose their numbness."

Although he ate well, France lost 17 pounds. The trip was so taxing that the men ate great quantities of fat to replace their diminishing body reserves. They added chunks of margarine to soups and piled it an inch thick on their bread. Even so, they were constantly hungry during the last month.

The journey was stressful, but France experienced a sense of danger only once. In June the men were caught in a four-day storm on an ice cap, with white-out conditions and winds up to 40 miles an hour. "We could barely stand up and if we had separated, we wouldn't have found each other."

Ellesmere Island is home to only about 150 people and the four saw no one from the time they were dropped off to the time they arrived at their destination. "We had 96 days of solitude and by the end of the trip we had started to get on each other's nerves."

Now back in Toronto, France plans to write a book—as soon as he adjusts to city life. "It's difficult going from a timeless environment to one where everybody counts minutes and seconds. It was very much a 'know thyself' sort of trip. It was easy to objectify my life and look at my strengths and weaknesses and gain some sort of perspective. Ellesmere is like a mirror in which I could see myself unfettered by the trappings and shields of civilization."

Arctic No Longer Pristine, WWF Researcher Says — M. Chrumka, *Working for Wildlife*, World Wildlife Fund Canada, Winter 1991

As many of us endure the cold days of January and February, it is difficult to imagine why anyone would willingly spend three months trekking in the High Arctic. This past year, however, one of World Wildlife Fund Canada's project co-ordinators, Robert France, joined three others in a 1,300 kilometre journey across Canada's northern-most island. The intent of this expedition, entitled Arctic Light 1990, was to complete the first human-powered traverse of Ellesmere Island. France and the rest of the team—John Dunn of Calgary, Graeme Magor of Hamilton, Ontario and Mike Sharp of Vancouver—departed from Calgary on a balmy day last March and began their trek at Cape Columbia in minus 30 degree weather. After 96 days, the journey ended in early July at Grise Fiord.

An environmental scientist from the University of Toronto, France leapt at the opportunity to participate in the expedition and conduct a study examining the impact of the long-range transportation of air pollutants. This project was supported by WWF through the Wildlife Toxicology Fund.

During the course of the trek, France collected lichen samples and now back in Toronto, is farming out the samples for analysis. France expects to find quantities of pesticides, industrial mercury, concentrations of sulphur emitted from power plants and lead emissions from cars in his lichen samples. He hopes to determine the extent to which air pollutants are infiltrating the "pristine" regions of the globe, as well as consider the potential for transfer of these pollutants up the food chain.

Although the results of his sampling will not be available for a couple of months, France was struck by the pollution which is already apparent in the Arctic. He maintains that the notion of an "untainted northern environment" is a myth. He saw many examples of human defacement of this wilderness area—surveying material discarded by cartographers, cigarette butts and coffee cups left behind by helicopter pilots after refuelling. France insists that a "garbage in, garbage out" policy must be enforced for the arctic.

As well as collecting samples and appreciating the amount of pollution that has already affected Ellesmere Island, France also conducted the first major wildlife survey of the island from the ground. He spotted, for example, a number of gyrfalcon, Musk Ox, Snowy Owl and several colonies of Ivory Gulls, a threatened species in Canada.

To date, his survey results are emphasizing the crucial role of polynyas in the Arctic. Polynyas are bodies of water kept permanently open by ocean currents. They attract a large variety of wildlife such as geese, duck and other birds, seals, walrus and polar bears. France was excited to discover that almost 80 percent of the wildlife he surveyed were concentrated around polynyas. He insists that these areas must remain as sanctuaries in order to sustain Arctic wildlife over broad regions. Oil exploration transportation within these bodies of water would undoubtedly be devastating.

Now that the expedition is over, France is eagerly awaiting the results of the analyses being done on his samples and is organizing his survey information for publication. Once these projects are complete, he hopes to write a popular account of the trip, which will undoubtedly contribute significantly to a greater understanding of our Arctic environment.

Mind and Matter Column: The Road to Eden is Lined With the Garbage of Well-Meaning Researchers — S. Strauss, *The Globe and Mail*, March 28, 1992

Let's talk garbage and man's immortal soul. Recently readers of the sober British science magazine Nature were confronted with the argument that detritus and humanity's fall from grace are inextricably linked.

And, miracle of miracles, the proof of man's sinfulness comes from a region of Eden that Eve's children call Canada.

What happened is that in 1990, McGill University biologist Robert France was part of a 1,200-km ski traverse of Ellesmere Island in the Northwest Territories. "We came to realize," wrote Dr. France, "that if curiosity in paradise was the original sin, then certainly the cavalier jettison of human garbage must be a close second."

While earlier explorers may have been dazzled by ice fields and polar bears, what France sway was garbage. He not only saw it, he assiduously counted it. There were 208 bits of human ejectus—47 abandoned gasoline drums, 41 rusted metal food containers, 15 rifle cartridges, 14 snowmobile parts, six cigarette butts and various other signs of mankind's passage. (Especially striking were the three-metre high words "blow me" trampled in the snow for the amusement of incoming helicopter pilots.)

Dr. France knew the name of the chief garbage sinners. Geographer John England of the University of Alberta and a U.S. compatriot left behind the biggest mess that Dr. France came across: a scattered cache of food and other supplies. The McGill scientist saw a certain irony in the fact that researchers studying paradise ended up polluting it.

But there was more. Wasn't this the same Professor England who had argued against the establishment of a national park at Ellesmere Island because "it would attract too many tourists to an 'environment [that] has exceptional limits in sustaining and suffering human use'"

A number of difficult truths flow from Dr. France's tale of Arctic garbage tourism. In an interview, Prof. England admitted to being red-faced about his mess, which since the 1990 trip has in fact been cleaned up by Parks Canada and charged against his research funding. Dr. England says he was underfunded in 1976 when the cache was laid down, but nonetheless he always intended to go back. However, he and others point out that the realities of doing research in the Arctic can involve hard choices. You leave material behind because it is often extremely difficult and expensive to have things flown out. Air transport costs about $1,000 and hour and you can find yourself having to bend plans to fit the whims of pilot schedules.

"A pilot will fly in and say 'I have enough room to take three people and their gear. What do you want to do?' Just leaving people there doesn't seem like an alternative," says Josef Svoboda, a botany professor at the University of Toronto.

Overriding all constraints on good housekeeping is a historical issue that can be slippery as a frozen pack. When and why does one man's garbage become another's historical artifact? The Parks Canada rule is that anything that is older than 40 years takes on an aura of ancient truth-telling.

"The artifacts at Fort Conger [the place where Robert Peary, the first person to reach the North Pole, and other early explorers and scientists camped] basically are piles of rusty tin cans. When I talk to people about the park I can show one slide of an old research camp that is 25 to 30 years old and say this is a national disgrace, then show Fort Conger and because it is over 100 years old say it is a national treasure," says Renee Wissink, senior park warden for Ellesmere National Park Reserve.

To clean up the eyesore, Parks Canada has carried away 25,000 kg of refuse from research sites set up in 1956-57 for International Geophysical Year. At the same time, it has set aside and stored relics of history that were moldering there. The artifacts are not the gold of Troy but the plastic and iron of evolving Arctic technology: belt-driven early Ski-Doos and primitive Caterpillar tractors.

The moral: Human ejecta may as easily define as pollute a place. Our modern fall from grace is not what we leave behind, but how much.

Junk of the North — Anon., *BBC Wildlife Magazine*, April 1992

Ellesmere Island in the Canadian High Arctic, home of white wolves and musk ox, is littered with the rubbish and bric-a-brac of two centuries of exercises and expeditions—including those of notable adventurers, such as Sir Ranulph Finnes—according to Dr. Robert France, of the recent Arctic Light Expedition, financed by WWF Canada and the Royal Canadian Geographic Society. The expedition's survey highlights, according to Dr. France, that "though some of the most remote corners of the globe may be…classified as wilderness…nowhere is the world really free of human waste."

The most common item of rubbish on the island is the gasoline drum, left behind by the aircraft which pick up returning scientific expeditions. The solution, says Dr. France, is easy: collapsible plastic 'fuel bladders', as used in the Antarctic, which when empty, can be folded up and removed.

Scientific equipment taken on to the island should be recorded on a ledger dept at the main base in Resolute, and checked off when scientists return. Toilet paper should be burned, and other garbage cached and later removed.

As far as the "noxious, environmentally unsound activities" of the Canadian military are concerned, there is no need or them in such a sensitive Arctic environment—particularly with the recent demise of the Soviet Union. But in the case of graffiti, says Dr. France, "the solution…is beyond fathoming."

Trivializes efforts of Arctic researchers — John England, *The Globe and Mail,* April 11, 1992

A column by Stephen Struass, The Road to Eden is Lined with the Garbage of Well-meaning Researchers, misrepresents Arctic researchers in an alarming and manipulative manner (*Focus*–March 28). The article is based on the observations of Dr. Robert France, a biologist at McGill University, who counted "208 bits of human ejecta" along a 1,200-kilometre transect of Ellesmere Island back in 1990.

Dr. France's moralizing does considerable disservice to numerous, conscientious Arctic scientists who have organized field work under difficult conditions and modest funding. These individuals have made Canada a world leader in understanding Arctic environments and Dr. France has publicly trivialized this contribution.

So, in midst of our constitutional wrangling it is appropriate that England finally responds to France.

Mr. Strauss dramatically identifies me and my American colleague as the "chief garbage sinners" who left behind a "Scattered cache of food and other supplies." The cache was left in 1976 secured in metal and wooden boxes for future use. This was common practice because it is prohibitively expensive to ship these supplies back south in aircraft that cost more than a thousand dollars per hour! What Dr. France naively called "garbage" were essential supplies in this isolated area. That the cache was scattered when Dr. France arrived suggests that someone had tampered with it. My freight canoe also disappeared.

Mr. Strauss carelessly states that I was "red-faced" about this incident. Naturally I was disappointed that this had occurred but there was nothing that would warrant my embarrassment. Because the remainder of the cache was removed over a year ago (solving the problem) the reader can only conclude that Dr. France's finger-pointing is inane. He is beating a dead horse.

Mr. Strauss also insinuates that, 10 years ago, I was trying to exclude tourists while preserving a private sanctuary for Arctic scientists. This conclusion does not represent my argument that the environment could sustain only minimal impact. I have always accepted that the constraints deemed necessary for conservation be applied equally to tourists and scientists. The column also gives the reader the impression that Arctic scientists have an arrogant and cavalier attitude toward the environment. I have worked in the Arctic for more than 25 years and my graduate students and I have had hundreds of camps. We have always made the effort to take what we brought in, down to the last shreds of string. If Dr. France would like to make a career out of playing Sherlock Holmes, I would gladly point him in the right direction. On numerous occasions we have picked up debris left by others.

If Dr. France wants to alert the public about Arctic pollution he would be better off not dwelling on cigarette butts and temporary graffiti trampled in the snow, but rather concentrate on other issues such as the toxic "Arctic haze" that invades the atmosphere each winter.

If one stops and thinks about how little debris Dr. France actually encountered (not to condone it), I can only conclude that Dr. France was in paradise but simply failed to realize it! Through this kind of opportunistic journalism it is unfortunate that Mr. Strauss and Dr. France have chosen to add even more garbage to Eden.

Misrepresenting paradise — John England, *Nature*, June 4, 1992

SIR—I am the unidentified 'geologist' accused of despoiling parts of Ellesmere Island with "the largest collection of garbage" that Dr. France encountered on a ski-traverse1. France quotes one of my papers concerning conservation of the high Arctic and implies that my intention was selectively to discourage tourism because of the area's exceptional environmental sensitivity. On both issues France has misled Nature's readership.

First, the "garbage" to which France alludes is a former cache that was left in the field secured in metal and wooden boxes in 1976. Limited financial support precluded an extra aircraft flight that summer to remove it and our plan was to reuse this equipment. The cache was not abandoned, and what was garbage to France was sustenance to me! Until 1984, colleagues of mine worked at this site and used my freight canoe. The cache was revisited as recently as 1988 and was still secure, but the boat had disappeared, possibly stolen. In any event, the cache was removed to Resolute bay more than a year ago, so I fail to see the point of France's argument.

My articles about the preservation of the high Arctic environment never implied that tourists should be excluded while scientists enjoyed the area as a private sanctuary. I have worked on Ellesmere Island for more than 20 years and, together with my graduate students, we have had several hundred camps. All the times we have taken in have gone out, down to the last pieces of string.

Finally, France unwittingly reveals his unfamiliarity with Arctic research when he says that the Polar Continental Shelf Project should have a 'garbage in garbage out' policy. It has been doing so for more than a decade. France's naivety about Arctic biology is similarly revealed in his statement that the land has only "recently emerged from its protective cradle of ice". This ignores all the research that has demonstrated that large ice-free areas served as biological refugia for tens of thousands of years2-5.

Most fieldwork in the Arctic is conducted by conscientious researchers who cherish this environment. It is careless and misleading to imply that the Arctic if 'riddled' with their garbage which serves as testimony to their 'arrogance'. That France encountered only 208 objects across 1,200 km (six of them being cigarette butts) affirms that he was in paradise but simply failed to recognize it.

[1] *France, R. Nature 355:504 (1992)*

[2] *England, J. & Bradley, R.S. Science 200, 265-270 (1978)*

[3] *England, J. Can Geog. 103, 8-17 (1983)*

[4] *Stewart, T.G. & England, J. Boreas 15, 25-31 (1986)*

[5] *Lemmen, D.S. Can. J. Earth Sci. 26, 2578-2590 (1989)*

Literature Cited

Note: Citation formatting varied among the different journals in which the original papers were published.

Addison, R.F. and M.E. Zinck. 1986. *Environ. Sci. Tech.* 20:253-256.

Atlan, H. 1985. Information theory and self-organization in ecosystems. pg. 187-199 *In* Ulanowicz and T. Platt (Eds.) Ecosystem theory for biological oceanography. *Can. Bull. Fish. Aquat. Sci.* No. 213.

Ambach, W. and F. Muller. 1980. Determination of net accumulations from gross Beta activity measurements in the North Water region. *Polarforschung* 50:1-7.

Ault, W.U., R.G. Senechel and W.F. Eriebach. 1970. Isotopic compositions as a natural tracer of lead in the environment. *Environ. Sci. Tech.* 6:705-708.

Bacci, E., D. Calamari, C. Gaggi, R. Fanelli and M. Morosini. 1986. *Chemosphere* 15:747-754.

Barrie, L.A. 1986. *Atm. Environ.* 20:643.

Barrie, L.A., R.M. Hoff and S.M. Daggupaty. 1981. *Atm. Environ.* 15:1407.

Barrie, L.A., et al. 1992. Arctic contaminants: Sources, occurrence and pathways. *Sci. Total Environ.* 122:1-24.

Barry, T.W. 1968. Observations on natural mortality and native use of eider ducks along the Beaufort Sea coast. *Can. Field Nat.* 82:140-144.

Baskaran, M., J.J. Kelley, A.S. Naidu and D.F. Holleman. 1991. Environmental radiocesium in subarctic and arctic Alaska following Chernobyl. *Arctic* 44:346-350.

Bateson and Plowright. 1959. *Brit. Birds* 59:105-114.

Bengston, S.A. 1971. *Ibis* 113:468-476.

Benton, T. 1989. *Nature.* 352:113.

Bidleman, T.F. et al. 1989. *Arctic* 42:307-313.

Bird, P.M. 1968. Studies of fallout 137Cs in the Canadian North. *Arch. Environ. Health.* 17:631-638.

Blais, J.M. 1996. Using isotopic tracers in lake sediments to assess atmospheric transport of lead in eastern Canada. *Wat., Air Soil Pollut.* 92:329-342.

Blomqvist, S. and M. Elander. 1981. Sabine's gull (Xema sabini), Ross's gull (Rhodostethia rosea) and Ivory gulls (Pagophilia eburnean) in the Arctic: A review. *Arctic* 34:122-132.

Boutron, C.F., U. Gorlach, J.P. Candelone, M.A. Bolshov and R.J. Delmos. 1991. Decrease in anthropogenic lead, cadmium and zinc in Greenland snows since the late 1960s. *Nature* 353:153-156.

Bray, R. 1943. Notes on the birds of Southampton Island, Baffin Island and Melville Peninsula. *Auk* 60:504-536.

Brown, R.G. and D.N. Nettleship. 1981. The biological significance of polynyas to arctic colonial seabirds. pg. 59-65. In Stirling, I and H. Cleator (Eds.) Polynyas in the Canadian Arctic. *Can. Wild. Serv.* Occasional Paper No. 45.

Brown, R.G., T. Davis and D.N. Nettleship. 1974. Ivory gulls, Pagophila eburnean, on the water. *Can. Field Nat.* 88:368-369.

Burnham, W.A. and W.G. Mattox. 1984. Biology of the peregrine and gyrfalcon in Greenland. Meddelei serom Gronland. *Bioscience* 14:25 pg.

Calamari, D. et al. 1991. *Environ. Sci. Tech.* 25:1489-1495.

Calamari, D. P. Tremolada, A. Di Guardo and M. Vighi. 1994. *Environ. Sci. Tech.* 28:429-434.

Cameron, M. and I.M. Weis. 1993. *Arctic* 46:42-46.

Carlberg, G.E., E.B. Ofstad, H. Drangsholt and E. Steinnes. 1983. *Chemosphere* 12:341-356.

Caughley, G. 1974. Biases in aerial survey. J. *Wild. Manag.* 38:921-933.

Church, T.M. et al. 1990. Trace elements in the North Atlantic trophosphere: Shipboard results of precipitation and aerosols. *Global Biogeochem.* Cycl. 4:431-443.

Conroy, M.J., J.R. Goldsberry, J.E. Hines and R.S. Stott. 1988. Evaluation of aerial transect surveys for wintering American black ducks. J. *Wild. Manag.* 2:694-703.

Corbet, P.S. 1969. Terrestrial microclimate: Amelioration at high latitudes. *Science* 166:865-866.

Dalgety. 1932. *Brit. Birds* 26:2-7.

Dementiev, G.P. and N.N. Gortchakovskaya. 1945. *Ibis* 87:559-565.

Dick, L. 2001. Muskox land: Ellesmere Island in the age of contact. *Univ. Calgary Press.*

Diem, K.L. and K.H. Luk. 1980. Factors influencing waterfowl censuses in the parklands. J. *Wild. Manag.* 24:115-133.

Divoky, G.J. 1976. The pelagic feeding habits of Ivory and Ross's gulls. *Condor* 78:85-90.

Douglas, M.S., J.P. Smol and W. Blake. 1994. Marked post-18th century environmental change in High Arctic ecosystems. *Science* 266:416-419.

Dunbar, M. 1981. Physical causes and biological significance of polynyas and other open water in sea ice. pg. 29-45 In Stirling, I and H. Cleator (Eds.) Polynyas in the Canadian Arctic. *Can. Wild. Serv.* Occasional Paper No. 45.

Dunbar, M. and M.J. Dunbar. 1972. The history of the North Water. *Proc. Royal Soc.* (B) 72:231-241.

Dunn, J. 1990. Odyssey in the High Arctic; The Trans-Ellesmere Island Expedition. *Can. Geog.* 110:55-64.

Dunn, J. 1991. Arctic light: One hundred days on Ellesmere Island. *Explore* 50:26-30, Spring 1991.

Dunn, J. and G. Magor. 1989. Ellesmere Island: The land, human history, flora and fauna. Funding brochure.

Ehrenfield, D. (ed.) 1995. To preserve biodiversity- an overview: Readings in Conservation Biology. *Blackwell Sci. Inc.*

England, J. 1983. Ellesmere Island needs special attention. *Can. Geog.* 103:8-17.

Environment Canada. 1989. Ellesmere Island National Park Reserve. Interim. Manag. Guidelines. *Environ. Can.*, Can. Parks Serv.

Evans, R.D. and F.H. Rigler. 1985. *Wat., Air Soil Pollut.* 24:141.

Feely, H.W., H.L. Volchuk, E.P. Hardy and L.E. Toonkel. 1978. *Environ. Sci. Tech.* 12:808-809.

Fellin, P. et al. 1996. *Environ. Toxicol. Chem.* 15:253-261.

Fielden, H.W. 1877a. On the Mammalia of North Greenland and Grinnell Land. *Proc. Zool. Soc.* Ser. 3, Vol. 1.

Fielden, H.W. 1877b. List of birds observed in Smith Sound and the Polar Basin, during the Arctic Expedition of 1875-76. *Ibis* 1:401-412.

Finley, K.J. and W.E. Renaud. 1980. Marine mammals inhabiting the Baffin Bay North Water in winter. *Arctic* 33:231-241.

Focardi, S., C. Gaggi, C. Chemelio and E. Bacci. 1991. *Polar Record* 27:241.

Folkeson, L. 1979. *Wat., Air Soil Pollut.* 24:241-

France, R. L. 2003. *Wetland design: Principles and practices for landscape architects and land-use planners.* W.W. Norton

France, R.L. 1990. Theoretical framework for developing and operationalizing an index of zoobenthos community integrity: Application to biomonitoring with zoobenthos communities in the Great Lakes. pg. 169-194 In Edwards, C.J. and H.A. Regier (Eds.) An ecosystem approach to the integrity of the Great Lakes in turbulent times. *Great Lakes Fish. Comm. Special Publ.* 90-4.

France, R.L. 2006. *Introduction to watershed development: Perspectives to aid in managing sprawl.* Rowman and Littlefield.

Fraser, J.K. 1957. Birds observed in the central Canadian Arctic, 1953, 1955, 1956. *Can. Field Nat.* 71:192-199.

Freedman, B. and J. Svoboda. 1982. *Can. Field Nat.* 96:56-60.

Frisch, T. 1983. Ivory gull colonies on the Devon Island Ice Cap, arctic Canada. *Arctic* 36:370-371.

Frisch, T. and W.C. Morgan. 1979. Ivory gull colonies in southeastern Ellesmere Island, arctic Canada. *Can. Field Nat.* 93:173-174.

Gaggi, C., E. Bacci, D. Calamari and R. Fanelli. 1985. *Chemosphere* 14:1673-1686.

Gasaway, W.C., S.D. Dubois and S.J. Harbo. 1985. Biases in aerial transect surveys for moose during May and June. J. *Wild. Manag.* 49:777-784.

Glooschenko, W.A. and J.A. Capobianco. 1978. *Wat., Air Soil Pollut.* 11:253.

Gobeil, C., W.K. Johnson, R.W. Macdonald and C.S. Wong. 1995. Sources and burdens of lead in St. Lawrence estuary sediments: Isotopic evidence. *Environ. Sci. Tech.* 29:193-203.

Godfrey, W.E. 1953. Notes on Ellesmere Island birds. *Can. Field Nat.* 67:192-199.

Godfrey, W.E. 1966. *The birds of Canada.* Nat. Mus. Nat. Sci.

Godfrey, W.E. 1986. *The birds of Canada. Revised Edition.* Nat. Mus. Nat. Res.

Gorham, E. 1959. A comparison of lower and higher plants as accumulators of radioactive fallout. Can. J. Bot. 37:327-329.

Gould, J. 1988. A comparison of avian and mammalian fauna at Lake Hazen, Northwest Territories in 1961-62 and 1981-82. *Can. Field Nat.* 102:666-670.

Greely, A.W. 1886. *Three years of arctic service, an account of the Lady Franklin Bay Expedition of 1881-84 and the attainment of the farthest north.* 2 vols. Scribner's.

Gregor, D.J. and W.D. Gummer. 1989. *Environ. Sci. Tech.* 23:561-565.

Groet, S.S. 1976. *Oikos* 27:445.

Haig-Thomas, D. 1939. *Tracks in the snow.* Hudder and Stroughten.

Hanson, W.C. 1967. Cesium-137 in Alaskan lichens, caribou and Eskimos. *Health Phys.* 13:383-389.

Hanson, W.C. 1968. Fallout radionuclide distributions in Northern Alaskan ecosystems. *Arch. Environ. Health* 17:639-648.

Hanson, W.C. 1971. Fallout radionuclide distributions in lichen communities near Thule. *Arctic* 24:269-276.

Hanson, W.C., and L.L. Eberhardt. 1971. Cycling and compartmentalizing of radionuclides in northern Alaskan lichen communities. pg. 71-75 *In* Nelson, D.J. (Ed.) *Radionuclides in ecosystems. Proc. Third Nat. Symp.* Oak Ridge, Tenn., Nat. Tech. Info. Serv.

Hardy, E.P., M. Meyer, J.S. Allen and L.T. Alexander. 1968. Strontium-90 on the Earth's surface. *Nature* 219:584-587.

Hargrave, B.T., W.P. Vass, P.E. Erickson and B.R. Fowler. 1988. *Tellus* 40B:480-493.

Hoff, R.M. and K.W. Chan. 1986. *Chemosphere* 15:449-452.

Hopper, J.F. and L.A. Barrie. 1988. Regional and background aerosol trace elemental composition observed in eastern Canada. *Tellus* 40B:446-462.

Hopper, J.F., H.B. Ross, W.T. Sturges and L.A. Barrie. 1991. Regional source discrimination of atmospheric aerosols in Europe using the isotopic composition of lead. *Tellus* 43B:45-60

Hussell, D.J. and G.L. Holroyd. 1974. Birds of Truelove Lowland and adjacent areas of northwestern Devon Island, NWT. *Can. Field. Nat.* 88:197-212.

Hutchison-Benson, E., J. Svoboda and H.W. Taylor. 1985. The latitudinal inventory of ^{137}Cs in vegetation and topsoil in northern Canada. *Can. J. Bot.* 63:784-791.

Jenkins, M.A. 1978. *Auk* 95:122-127.

Johansson, K. 1989. *Wat., Air Soil Pollut.* 47:441.

Joshi, S.R. 1987. Early Canadian results on the long-range transport of Chernobyl radioactivity. *Sci. Total Environ.* 63:125-137.

Karr, J.R. 1990. Biological integrity and the goal of environmental legislation: Lessons for conservation biology. *Conserv. Biol.* 4:244-250.

Kay, J. J. 1990. A non-equilibrium thermodynamic framework for discussing ecosystem integrity. pg. 209-238 *In* Edwards, C.J. and H.A. Regier (Eds.) *An ecosystem approach to the integrity of the Great Lakes in turbulent times.* Great Lakes Fish. Comm. Special Publ. 90-4.

Kidd, K.A., D.W. Schindler, R.H. Hesslien and D.C. Muir. 1995. *Sci. Total Environ.* 160/161:381-390.

Kiliaan, H.P. and I. Stirling. 1978. Observations on overwintering walruses in the eastern Canadian High Arctic. J. *Mammal.* 59:197-200.

Kobalenko, J. 2002. *The horizontal Everest: Extreme journeys on Ellesmere Island.* Soho Press.

Koranda, J.J. and J.R. Martin. 1971. Gamma-emitting radionuclides in Alaskan environments 1967-1970. pg. 81-97 *In* Nelson, D.J. (Ed.) *Radionuclides in ecosystems.* Proc. Third Nat. Symp. Oak Ridge, Tenn., Nat. Tech. Info. Serv.

Larson. 1960. *Oikos* 11:276-305.

Lead, W.A., E. Steinnes and K.C. Jones. 1996. *Environ. Sci. Tech.* 30:524-530.

Lopez, B. 1987. *Arctic dreams.* Bantam.

Lotz, J. 2006. *The best journey in the world: Adventures in Canada's High Artic.* Pottersfield Press.

Lowenthal, D.H. and K.A. Rahn. 1985. *Atm. Environ.* 19:2011.

Luckas, B., W. Vetter, P. Fischer, G. Heidemann and J. Plotz. 1990. *Environ. Sci. Tech.* 28:13-19.

MacDonald, S.D. 1953. *Report of biological investigations at Alert, NWT.* Nat. Mus. Can. Bull. 128:241-256.

MacDonald, S.D. 1976. Phantoms of the polar ice. *Audubon* 78:2-19.

MacMillian, D.B. 1918. Four years in the white north. *Harper and Brothers.*

Magor, G. 1991. Beneath the Arctic light: A personal view of the crossing of Ellesmere Island. *Hamilton This Month* VII/5. May Issue.

Mattson, S. and K. Liden. 1975. ^{137}Cs in carpets of the forest moss Pleurozium screberi 1961-1973. *Oikos* 26:3232-327.

McLaren, P.L. 1982. Spring migration and habitat use by seabirds in Eastern Lancaster Sound and western Baffin Bay. *Arctic* 35:88-111.

McLaren, P.L. and M.A. McLaren. 1982. Waterfowl populations in eastern Lancaster Sound and western Baffin Bay. *Arctic* 35:149-157.

McLaren, P.L. and W.E. Renaud. 1982. Seabird concentrations in late summer along the coasts of Devon and Ellesmere Islands, NWT. *Arctic* 35:112-117.

McNeely, R. and W.D. Gummer. 1984. *Arctic* 37:210-233.

Mohindra, V.K. 1967. Cesium-137 burdens in the Canadian north. Acta Radiol. *Theor. Phys. Biol.* 6:481-490.

Muir, D.G.C., R.J. Norstrom and M. Simon. 1988. *Environ. Sci. Tech.* 22:1071-1079.

Muir, D. G., C.A. Ford, N.P. Grift, D.A. Metner and W.L. Lockhart. 1990. *Arch. Environ. Toxicol.* 19:530-542.

Muir, D. G., C.A. Ford, N.P. Grift, D.A. Metner and W.L. Lockhart. 1990. *Environ. Contam. Toxicol.* 27:1201-1210.

Muir, D.G. et al. 1995. *Sci. Total Environ.* 160/161:447-457.

Muir, D.G. et al. 1996. *Environ. Sci. Tech.* 30:3609-3617.

Muir, D. et al. 1996. Ecosystem uptake and effects. Chapter 3 *In* Shearer, R. (Ed.) *Canadian arctic contamination assessment report.* Dept. Indian and North. Affairs.

Murozumi, M., T.J. Chow and C.C. Patterson. 1969. Geochim. *Cosmochim. Acta* A33:1247.

Nettleship, D.N. and W.J. Mahr. 1973. The avifauna of Hazen Camp, Ellesmere Island, NWT. *Polarforschung* 44:66-74.

Nettleship, D.N. and P.A. Smith. (Eds.) 1975. Site 2-3: Lake Hazen, Ellesmere Island. pg. 62-70 *In Ecological sites in northern Canada.* Can. Comm. Int. Biol. Prog., Conserv. Terrest., Panel 9.

Nettleship, D.N., T.R. Birkhead and A.J. Gaston. 1980. Reproductive failure among arctic seabirds associated with unusual ice conditions in Lancaster Sound. *Can. Wild. Serv. Manual* Rep. 77.

Nieboer, E., H.M. Ahmed, K.J. Puckett and D.H. Richardson. 1972. *Lichenol.* 5:292.

Norstrom, R.J., M. Simon, D.C. Muir and R.E. Schweinburg. 1988. *Environ. Sci. Tech.* 22:1063-1071.

Noss, R.F. and L.D. Harris. 1986. Nodes, networks, and MUMs: Preserving diversity at all scales. *Environ. Manag.* 10:299-309.

Nriagu, J.O. and J.M. Pacyna. 1988. *Nature* 333:134.

Nutt, D.C. 1969. The North Water of Baffin Bay. *Polar Notes* 9:1-25.

Oehme, M. 1992. *Ambio* 20:293-297.

Orr, C.D. and J.L. Parsons. 1982. Ivory gulls, Pagophila eburnean, and ice edges in Davis Strait and the Labrador Sea. *Can. Field Nat.* 96:323-328.

Ouelett, M. and H.G. Jones. 1983.*Can. J. Earth Sci.* 20:23.

Pakarinen, P. and K. Tolonen. 1976. *Ambio* 5:38.

Pakarinen, P. and K. Tolonen. 1977. *Oikos* 28:69.

Patton, G.W.., M.D. Walla, T.F. Bidleman and L.A. Barrie. 1989. *Tellus* 41B:243-255.

Parmelle, D.F. and S.D. MacDonald. 1960. *The birds of westcentral Ellesmere Island and adjacent areas.* Nat. Mus. Can. Bull. 169.

Pattie, D.L. 1990. A 16-year record of summer birds on Truelove Lowland, Devon Island, Northwest Territories, Canada. *Arctic* 43:275-283.

Patton, G.W., D.A. Hinckley, M.D. Walla and T.F. Bidleman. *Tellus* 41B:243-255.

Patton, G.W., M.D. Walla, T.F. Bidleman and L.A. Barrie. 1991. *Geophys.* Res. 96:10867-10877.

Peters, R. 1997. *Publication and citation rankings of prominent ecologists.* Unpubl. MS.

Poole, K.G. and G.A. Boag. 1988. Can. J. Zool. 66:334-344.

Porsild, A.E. 1951. Bird notes from Banks and Victoria islands. *Can. Field Nat.* 65:40-42.

Prach, R.W., M.H. Boyd and F.G. Cooch. 1981. Polynyas and sea ducks. pg. 67-70 In Stirling, I and H. Cleator (Eds.) Polynyas in the Canadian Arctic. *Can. Wild. Serv.* Occasional Paper No. 45.

Puckett, K.J. and E.J. Finegan. 1977. *The use of multivariate techniques in the analysis of metal levels found in lichens.* Atm. Environ. Serv., Environ. Can.

Puckett, K.J. and E.J. Finegan. 1980. Can. J. Bot. 58:2073.

Rabinowitz, M.B. and G.W. Witherill. 1972. Identifying sources of lead contamination by stable isotope techniques. *Environ. Sci. Tech.* 6:705-708.

Rahn, K.A. 1981. *Atm. Environ.* 15:1457.

Rahn, K.A. 1981. *Atm. Environ.* 15:1447.

Reed, C. 2007. "An oracle of agua": Immersed in his own element. *Harvard Mag.* Jan./Feb. Issue

Reed, A. and P. Dupuis. 1983. Ivory gulls, Pagophila eburnean, nesting on the Brodeur Peninsula, Baffin Island, N.W.T. *Can. Field Nat.* 97:332.

Renaud, W.E. and M.S. Bradstreet. 1980. Late winter distribution of Black Guillemots in northern Baffin Bay and the Canadian High Arctic. *Can. Field Nat.* 94:421-425.

Renaud, W.E. and P.L. McLaren. 1982. Ivory gull (Pagophila eburnean) distribution in late summer and autumn in eastern Lancaster Sound and western Baffin Bay. *Arctic* 35:141-148.

Rickard, W.H., J.J. Davis, W.C. Hanson and D.G. Watson. 1965. Gamma-emitting radionuclides in Alaskan tundra vegetation. *Ecology* 46:352-356.

Robbins, C.E., B. Briuin and H.S. Zim. 1989. *A guide to field identification of birds of North America.* Golden Press.

Rosman, K.J., W. Chisholm, C.F. Boutron, J.P. Canelone and U. Gorlach. 1993. Isotopic evidence for the source of lead in Greenland snows since the late 1960s. *Nature* 362:333-335.

Ross, H.B. 1987. Trace metals in precipitation in Sweden. *Wat., Air Soil Pollut.* 36:349-356.

Ross, H.B. 1990. *Wat., Air Soil Pollut.* 50:63.

Rowan, D.J. and J.B. Rasmussen. 1994. *Can. J. Fish. Aquat. Sci.* 51:2388-2410.

Rowan, D.J. and J.B. Rasmussen. 1995. *J. Appl. Ecol.* 32:739-744.

Ruhling, A. and G. Tyler. 1971. *J. Appl. Ecol.* 8:497.

Ruhling, A. and G. Tyler. 1973. *Wat., Air Soil Pollut.* 2:445.

Sadler, H.E. 1974. On a polynya in Makinson Inlet. *Arctic* 27:157-159.

Saville, D.B. 1964. General ecology and vascular plants of the Hazen Camp area. *Arctic* 17:237-258.

Saville, D.B. and D.R. Oliver. 1964. Bird and mammal observations at Hazen Camp, northern Ellesmere Island. *Can. Field Nat.* 78:1-7.

Schledermann, P. 1980. Polynyas and prehistoric settlement patterns. *Arctic* 33:292-302.

Schledermann, P. 1996. *Voices in stone: A personal journey into the Arctic Past.* Arctic Inst. N. America.

Shirahata, H., R.W. Elia, C.C. Patterson and M. Koide. 1980. Chronological variations in concentrations and isotopic compositions of anthropogenic atmospheric lead in sediments of a remote subalpline pond. Geochem. *Acta* 44:149-162.

Simonich, S.L. and R.A. Hites. 1995. *Science* 269:1851-1854.

Smith, D.R., S. Niemeyer, J.A. Estes and A.R. Flegal. 1990. Stable lead isotopes evidence anthropogenic contamination in Alaskan sea otters. *Environ. Sci. Tech.* 24:1317-1321.

Smith, M. and B. Rigby. 1981. Distribution of polynyas in the Canadian Arctic. pg. 7-28 *In* Stirling, I and H. Cleator (Eds.) *Polynyas in the Canadian Arctic.* Can. Wild. Serv. Occasional Paper No. 45.

Snell, R.R. 1989. Status of Larus gulls at Home Bay, Baffin Island. *Colonial Waterbirds* 12:12-23.

Soper, J.D. 1940. Local distribution of eastern Canadian arctic birds. *Auk* 57:13-21.

Soper, J.D. and J.M. Powell. 1985. *Botanical studies in the Lake Hazen Region, northern Ellesmere Island, Northwest Territories, Canada.* Nat. Mus. Can.

Sordhal, S.A. 1990. *Wilson Bull.* 102:349-353.

Steedman, R.J. and H.A. Regier. 1987. Ecosystem science for the Great Lakes: Perspectives on degradative and rehabilitative transformations. *Can. J. Fish. Aquat. Sci.* 44 (Suppl. 2):95-103.

Steedman, R.J. and H.A. Regier. 1990. Ecological bases for an understanding of ecosystem integrity in the Great Lakes Basin. pg. 257-284 *In* Edwards, C.J. and H.A. Regier (Eds.) *An ecosystem approach to the integrity of the Great Lakes in turbulent times.* Great Lakes Fish. Comm. Special Publ. 90-4.

Sterling, I. 1980. The biological significance of polynyas in the Canadian Arctic. *Arctic* 33:303-315.

Sterling, I., H. Cleator and T. Smith. 1981. Marine mammals. pg. 45-58 *In* Stirling, I and H. Cleator (Eds.) *Polynyas in the Canadian Arctic.* Can. Wild. Serv. Occasional Paper No. 45.

Stewart, J. 1984. *Strategies of biomass allocation in some High Arctic saxifrages.* M.Sc. Thesis. Dalhousie Univ., Halifax.

Stott, R.S. and D.P. Olson. 1972. An evaluation of waterfowl surveys on the New Hampshire coastline. *J. Wild. Manag.* 36:468-477.

Sturges, W.T. and L.A. Barrie. 1987. Stable lead isotope ratios in arctic aerosols: Evidence for the origin of arctic air pollution. *Nature* 392:144-146.

Sturges, W.T. and L.A. Barrie. 1989. Stable lead isotope ratios in arctic aerosols: Evidence for the origin of arctic air pollution. *Atm. Environ.* 23:2513-2519.

Sturges, W.T., J.F. Hopper, L.A. Barrie and C. Schnell. 1993. Stable lead isotope ratios in Alaskan arctic aerosols. *Atm. Environ.* 27:2865-2871.

Sutherland, P.D. 1991. Archaeological site distributions on the south coast of Devon Island, High Arctic Canada. *Can. J. Archaeol.*

Sutton, G.M. and D.F. Parmelee. 1956. The rough-legged hawk in the American Arctic. *Arctic* 9:202-208.

Svoboda, J. and B. Freedman. 1981. *Ecology of a High Arctic lowland oasis at Alexandra Fiord (78° 53' N, 75° 55' W), Ellesmere Island, NWT, Canada.* Univ. Toronto, Dept. Botany.

Svoboda, J. and H.W. Taylor. 1979. Persistence of ^{137}Cs in arctic lichens, *Dryas integrifolia* and lake sediments. *Arct. Alp. Res.* 11:95-109.

Taylor, A. 1955. *Geographical discovery and exploration in the Queen Elizabeth Islands.* Can. Dept. Mines Tech. Surveys.

Taylor, H.W., E.A. Hutchison, K.L. McInnes and J. Svoboda. 1979. Cosmos 954: Search for airborne radioactivity on lichens in the crash area, Northern Territories, Canada. *Science* 205:1383-1385.

Taylor, H.W., E. Hutchison-Benson and J. Svoboda. 1985. Search for latitudinal trends in the effective half-life on fallout ^{137}Cs in vegetation of the Canadian Arctic. *Can. J. Bot.* 63:792-796.

Taylor, H.W., J. Svoboda, G.H. Henry and R.W. Wein. 1988. Post-Chernobyl ^{134}Cs levels at some localities in northern Canada. *Arctic* 41:293-296.

Tener, J.S. 1965. *Muskoxen in Canada; A biological and taxonomic review.* Can. Dept. N. Affairs and Nat. Res., Nat. Hist. Res. Branch, Can. Wild. Serv. Monogr. No. 2.

Thomas, V.G. and S.D. MacDonald. 1985. The significance of terrestrial and marine oases in the High Arctic. pg. 147-153 *In* Nelson, J.G., R. Needham and L. Norton (Eds.) *Proc. Arctic Heritage Symp.*, Banff, Alberta. Can. Parks Serv.

Thomas, V.G. and S.D. MacDonald. 1987. The breeding distribution and current population status of the Ivory gull in Canada. *Arctic* 40:211-218.

Thomas, W. 1984. *Wat., Air Soil Pollut.* 22:351-371.

Thomas, W. 1986. *Wat. Sci. Tech.* 18:47-57.

Ulanowicz, R.E. 1990. Ecosystem integrity and network theory. pg. 69-78 *In* Edwards, C.J. and H.A. Regier (Eds.) *An ecosystem approach to the integrity of the Great Lakes in turbulent times. Great Lakes Fish.* Comm. Special Publ. 90-4.

Vanderwaag, D.I. and C. Lamson. 1989. *The challenge of Arctic shipping: Science, environmental assessment and human values.* McGill-Queen's Univ. Press.

Villeneuve, J.P., E. Fogelqvist and C. Cattini. 1988. *Chemosphere* 17:399-403.

Villeneuve, J.P. and E. Holm. 1984. *Chemosphere* 13:1133-1138.

von Bertalanffy, L. 1950. The theory of open systems in physics and biology. *Science* 111:147-153.

von Droste, B. 1985. International approaches to Arctic heritage use and conservation. pg. 470-477 *In* Nelson, J.G., R. Needham and L. Norton (Eds.) *Proc. Arctic Heritage Symp.*, Banff, Alberta. Can. Parks Serv.

Walker, W. 1977. *Auk* 94:442-447.

Wania, F. and D. Mackay. 1993. *Ambio* 22:10-18.

Welch, H.E. et al. 1991. *Environ. Sci. Tech.* 25:280-286.

White, C.M. and T.J. Cade. 1971. *Living Bird* 10:107-150.

White, C.M. and R.B. Weiden. 1966. *Condor* 68:517-519.

Williams, S.R. 1980. *The report of the Joint Services Expedition to Princess Marie Bay, Ellesmere Island.* Defense Res. Bd., U.K.

Wooden, N. 1980. *Raptor Res.* 14:97-124.

Zakshek, E.M., K.J. Puckett and K.E. Percy. 1986. *Wat., Air Soil Pollut.* 30:161.

Scriptorium/Palimpsest Press

SPP

In medieval times, monk scholars would gather in the scriptorium and reuse valuable sheets of vellum parchment that had once contained previous words and images. In a similar spirit of conservation, Scriptorium/Palimpsest Press is devoted to collecting and republishing earlier palimpsest tracings.